Delmar's Test Preparation Series

Automobile Test

Automotive Parts Specialist (Test P2)

2nd Edition

DELMAR

THOMSON LEARNING ™

Australia   Canada   Mexico   Singapore   Spain   United Kingdom   United States

DELMAR

™

THOMSON LEARNING

Delmar's Test Preparation Series
Automobile Test

Automotive Parts Specialist (Test P2)
2nd Edition

Delmar Staff:

Business Unit Director:
Alar Elken

Executive Editor:
Sandy Clark

Acquisitions Editor:
Jack Erjavec

Team Assistant:
Bryan Viggiani

Developmental Editor:
Christopher Shortt

Executive Marketing Manager:
Maura Theriault

Executive Production Manager:
Mary Ellen Black

Production Manager:
Larry Main

Production Editor:
Betsy Hough

Production Editor:
Tom Stover

Channel Manager:
Mona Caron

Marketing Coordinator:
Brian McGrath

Cover Design:
Michael Egan

Cover Images Courtesy of:
DaimlerChrysler

NOTICE TO THE READER

Publisher does not warrant or guarantee any of the products described herein or perform any independent analysis in
connection with any of the product information contained herein. Publisher does not assume, and expressly disclaims, any
obligation to obtain and include information other than that provided to it by the manufacturer.

The reader is expressly warned to consider and adopt all safety precautions that might be indicated by the activities herein and to
avoid all potential hazards. By following the instructions contained herein, the reader willingly assumes all risks in connection
with such instructions.

The Publisher makes no representation or warranties of any kind, including but not limited to, the warranties of fitness for
particular purpose or merchantability, nor are any such representations implied with respect to the material set forth herein, and
the publisher takes no responsibility with respect to such material. The publisher shall not be liable for any special, consequential,
or exemplary damages resulting, in whole or part, from the readers' use of, or reliance upon, this material.

Contents

Section 5 Sample Test for Practice

Section 6 Additional Test Questions for Practice

Section 7 Appendices

Preface

This book is just one of a comprehensive series designed to prepare technicians to take and pass every ASE test. Delmar's series covers all of the Automotive tests A1 through A8 as well as Advanced Engine Performance L1 and Parts Specialist P2. The series also covers the five Collision Repair tests and the eight Medium/Heavy Duty truck tests.

Before any book in this series was written, Delmar staff met with and surveyed technicians and shop owners who have taken ASE tests and have used other preparatory materials. We found that they wanted, first and foremost, *lots* of practice tests and questions. Each book in our series contains a sample test and additional practice questions. You will be hard-pressed to find a test prep book with more questions for you to practice with. We have worked hard to ensure that these questions match the ASE style in types of questions, quantities, and level of difficulty.

Technicians also told us that they wanted to understand the ASE test and to have practical information about what they should expect. We have provided that as well, including a history of ASE and a section devoted to helping the technician "Take and Pass Every ASE Test" with case studies, test-taking strategies, and test formats.

Finally, techs wanted refresher information and references. Each of our books includes an overview section that is referenced to the task list. The complete task lists for each test appear in each book for the user's reference. There is also a complete glossary of terms for each booklet.

So whether you're looking for a sample test and a few extra questions to practice with or a complete introduction to ASE testing, with support for preparing thoroughly, this book series is an excellent answer.

We hope you benefit from this book and that you pass every ASE test you take!

Your comments, both positive and negative, are certainly encouraged! Please contact us at:

Automotive Editor
Delmar Publishers
3 Columbia Circle
Box 15015
Albany, NY 12212-5015

The History of ASE

History

Originally known as The National Institute for Automotive Service Excellence (NIASE), today's ASE was founded in 1972 as a non-profit, independent entity dedicated to improving the quality of automotive service and repair through the voluntary testing and certification of automotive technicians. Until that time, consumers had no way of distinguishing between competent and incompetent automotive mechanics. In the mid-1960s and early 1970s, efforts were made by several automotive industry affiliated associations to respond to this need. Though the associations were non-profit, many regarded certification test fees merely as a means of raising additional operating capital. Also, some associations, having a vested interest, produced test scores heavily weighted in the favor of its members.

From these efforts a new independent, non-profit association, the National Institute for Automotive Service Excellence (NIASE), was established. In early NIASE tests, Mechanic A, Mechanic B type questions were used. Over the years the trend has not changed, but in mid-1984 the term was changed to Technician A, Technician B to better emphasize sophistication of the skills needed to perform successfully in the modern motor vehicle industry. In certain tests the term used is Estimator A/B, Painter A/B, or Parts Specialist A/B. At about that same time, the logo was changed from "The Gear" to "The Blue Seal," and the organization adopted the acronym ASE for Automotive Service Excellence.

ASE

ASE's mission is to improve the quality of vehicle repair and service in the United States through the testing and certification of automotive repair technicians. Prospective candidates register for and take one or more of ASE's many exams.

Upon passing at least one exam and providing proof of two years of related work experience, the technician becomes ASE certified. A technician who passes a series of exams earns ASE Master Technician status. An automobile technician, for example, must pass eight exams for this recognition.

The exams, conducted twice a year at over seven hundred locations around the country, are administered by American College Testing (ACT). They stress real-world diagnostic and repair problems. Though a good knowledge of theory is helpful to the technician in answering many of the questions, there are no questions specifically on theory. Certification is valid for five years. To retain certification, the technician must be retested to renew his or her certificate.

The automotive consumer benefits because ASE certification is a valuable yardstick by which to measure the knowledge and skills of individual technicians, as well as their commitment to their chosen profession. It is also a tribute to the repair facility employing ASE certified technicians. ASE certified technicians are permitted to wear blue and white ASE shoulder insignia, referred to as the "Blue Seal of Excellence," and carry credentials listing their areas of expertise. Often employers display their technicians' credentials in the customer waiting area. Customers look for facilities that display ASE's Blue Seal of Excellence logo on outdoor signs, in the customer waiting area, in the telephone book (Yellow Pages), and in newspaper advertisements.

To become ASE certified, contact:

National Institute for Automotive Service Excellence
13505 Dulles Technology Drive
Herndon, VA 20171-3421

2 Take and Pass Every ASE Test

ASE Testing

Participating in an Automotive Service Excellence (ASE) voluntary certification program gives you a chance to show your customers that you have the "know-how" needed to work on today's modern vehicles. The ASE certification tests allow you to compare your skills and knowledge to the automotive service industry's standards for each specialty area.

If you are the "average" automotive technician taking this test, you are in your mid-thirties and have not attended school for about fifteen years. That means you probably have not taken a test in many years. Some of you, on the other hand, have attended college or taken postsecondary education courses and may be more familiar with taking tests and with test-taking strategies. There is, however, a difference in the ASE test you are preparing to take and the educational tests you may be accustomed to.

Who Writes the Questions?

The questions on all ASE tests are written by service industry experts familiar with all aspects of the subject area. ASE questions are entirely job-related and designed to test the skills that you need to know on the job.

The questions originate in an ASE "item-writing" workshop where service representatives from domestic and import automobile manufacturers, parts and equipment manufacturers, and vocational educators meet in a workshop setting to share their ideas and translate them into test questions. Each test question written by these experts is reviewed by all of the members of the group.

All of the questions are pretested and quality-checked in a nonscoring section of tests by a national sample of certifying technicians. The questions that meet ASE's high standards of accuracy and quality are then included in the scoring sections of future tests. Those questions that do not pass ASE's stringent test are sent back to the workshop or are discarded. ASE's tests are monitored by an independent proctor and are administered and machine-scored by an independent provider, American College Testing (ACT).

Objective Tests

A test is called an objective test if the same standards and conditions apply to everyone taking the test and there is only one correct answer to each question. Objective tests primarily measure your ability to recall information. A well-designed objective test can also test your ability to understand, analyze, interpret, and apply your knowledge. Objective tests include true-false, multiple choice, fill in the blank, and matching questions. ASE's tests consist exclusively of four-part multiple-choice objective questions.

Before beginning to take an objective test, quickly look over the test to determine the number of questions, but do not try to read through all of the questions. In an ASE test, there are usually between forty and eighty questions, depending on the subject. Read through each question before marking your answer. Answer the questions in the order they appear on the test. Leave the questions blank that you are not sure of and move on to the next question. You can return to those unanswered questions after you have finished the others. They may be easier to answer at a later time after your mind has had additional time to consider them on a subconscious level. In addition, you might find information in other questions that will help you to answer some of them.

Do not be obsessed by the apparent pattern of responses. For example, do not be influenced by a pattern like **d, c, b, a, d, c, b, a** on an ASE test.

There is also a lot of folk wisdom about taking objective tests. For example, there are those who would advise you to avoid response options that use certain words such as *all, none, always, never, must,* and *only,* to name a few. This, they claim, is because nothing in life is exclusive. They would advise you to choose response options that use words that allow for some exception, such as *sometimes, frequently, rarely, often, usually, seldom,* and *normally.* They would also advise you to avoid the first and last option (A and D) because test writers, they feel, are more comfortable if they put the correct answer in the middle (B and C) of the choices. Another recommendation often offered is to select the option that is either shorter or longer than the other three choices because it is more likely to be correct. Some would advise you to never change an answer since your first intuition is usually correct.

Although there may be a grain of truth in this folk wisdom, ASE test writers try to avoid them and so should you. There are just as many **A** answers as there are **B** answers, just as many **D** answers as **C** answers. As a matter of fact, ASE tries to balance the answers at about 25 percent per choice **A, B, C,** and **D.** There is no intention to use "tricky" words, such as outlined above. Put no credence in the opposing words "sometimes" and "never," for example.

Multiple-choice tests are sometimes challenging because there are often several choices that may seem possible, and it may be difficult to decide on the correct choice. The best strategy, in this case, is to first determine the correct answer before looking at the options. If you see the answer you decided on, you should still examine the options to make sure that none seem more correct than yours. If you do not know or are not sure of the answer, read each option very carefully and try to eliminate those options that you know to be wrong. That way, you can often arrive at the correct choice through a process of elimination.

If you have gone through all of the test and you still do not know the answer to some of the questions, then guess. Yes, guess. You then have at least a 25 percent chance of being correct. If you leave the question blank, you have no chance. In ASE tests, there is no penalty for being wrong.

Preparing for the Exam

The main reason we have included so many sample and practice questions in this guide is, simply, to help you learn what you know and what you don't know. We recommend that you work your way through each question in this book. Before doing this, carefully look through Section 3; it contains a description and explanation of the questions you'll find in an ASE exam.

Once you know what the questions will look like, move to the sample test. After you have answered one of the sample questions (Section 5), read the explanation (Section 7) to the answer for that question. If you don't feel you understand the reasoning for the correct answer, go back and read the overview (Section 4) for the task that is related to that question. If you still don't feel you have a solid understanding of the material, identify a good source of information on the topic, such as a textbook, and do some more studying.

After you have completed the sample test, move to the additional questions (Section 6). This time answer the questions as if you were taking an actual test. Once you have answered all of the questions, grade your results using the answer key in Section 7. For every question that you gave a wrong answer to, study the explanations to the answers and/or the overview of the related task areas.

Here are some basic guidelines to follow while preparing for the exam:

- Focus your studies on those areas you are weak in.
- Be honest with yourself while determining if you understand something.
- Study often but in short periods of time.
- Remove yourself from all distractions while studying.
- Keep in mind the goal of studying is not just to pass the exam, the real goal is to learn!

During the Test

Mark your bubble sheet clearly and accurately. One of the biggest problems an adult faces in test-taking, it seems, is in placing an answer in the correct spot on a bubble sheet. Make certain that you mark your answer for, say, question 21, in the space on the bubble sheet designated for the answer for question 21. A correct response in the wrong bubble will probably be wrong. Remember, the answer sheet is machine scored and can only "read" what you have bubbled in. Also, do not bubble in two answers for the same question.

If you finish answering all of the questions on a test ahead of time, go back and review the answers of those questions that you were not sure of. You can often catch careless errors by using the remaining time to review your answers.

At practically every test, some technicians will invariably finish ahead of time and turn their papers in long before the final call. Do not let them distract or intimidate you. Either they knew too little and could not finish the test, or they were very self-confident and thought they knew it all. Perhaps they were trying to impress the proctor or other technicians about how much they know. Often you may hear them later talking about the information they knew all the while but forgot to respond on their answer sheet.

It is not wise to use less than the total amount of time that you are allotted for a test. If there are any doubts, take the time for review. Any product can usually be made better with some additional effort. A test is no exception. It is not necessary to turn in your test paper until you are told to do so.

Your Test Results!

You can gain a better perspective about tests if you know and understand how they are scored. ASE's tests are scored by American College Testing (ACT), a non-partial, non-biased organization having no vested interest in ASE or in the automotive industry. Each question carries the same weight as any other question. For example, if there are fifty questions, each is worth 2 percent of the total score. The passing grade is 70 percent. That means you must correctly answer thirty-five of the fifty questions to pass the test.

The test results can tell you:
- where your knowledge equals or exceeds that needed for competent performance, or
- where you might need more preparation.

The test results *cannot* tell you:
- how you compare with other technicians, or
- how many questions you answered correctly.

Your ASE test score report will show the number of correct answers you got in each of the content areas. These numbers provide information about your performance in each area of the test. However, because there may be a different number of questions in each area of the test, a high percentage of correct answers in an area with few questions may not offset a low percentage in an area with many questions.

It may be noted that one does not "fail" an ASE test. The technician who does not pass is simply told "More Preparation Needed." Though large differences in percentages may indicate problem areas, it is important to consider how many questions were asked in each area. Since each test evaluates all phases of the work involved in a service specialty, you should be prepared in each area. A low score in one area could keep you from passing an entire test.

There is no such thing as average. You cannot determine your overall test score by adding the percentages given for each task area and dividing by the number of areas. It doesn't work that way because there generally are not the same number of questions in each task area. A task area with twenty questions, for example, counts more toward your total score than a task area with ten questions.

Your test report should give you a good picture of your results and a better understanding of your task areas of strength and weakness.

If you fail to pass the test, you may take it again at any time it is scheduled to be administered. You are the only one who will receive your test score. Test scores will not be given over the telephone by ASE nor will they be released to anyone without your written permission.

3 Types of Questions on an ASE Exam

ASE certification tests are often thought of as being tricky. They may seem to be tricky if you do not completely understand what is being asked. The following examples will help you recognize certain types of ASE questions and avoid common errors.

Each test is made up of forty to eighty multiple-choice questions. Multiple-choice questions are an efficient way to test knowledge. To answer them correctly, you must think about each choice as a possibility, and then choose the one that best answers the question. To do this, read each word of the question carefully. Do not assume you know what the question is about until you have finished reading it.

About 10 percent of the questions on an actual ASE exam will use an illustration. These drawings contain the information needed to correctly answer the question. The illustration must be studied carefully before attempting to answer the question. Often, techs look at the possible answers then try to match up the answers with the drawing. Always do the opposite; match the drawing to the answers. When the illustration is showing an electrical schematic or another system in detail, look over the system and try to figure out how the system works before you look at the question and the possible answers.

Multiple-Choice Questions

One type of multiple-choice question has three wrong answers and one correct answer. The wrong answers, however, may be almost correct, so be careful not to jump at the first answer that seems to be correct. If all the answers seem to be correct, choose the answer that is the most correct. If you readily know the answer, this kind of question does not present a problem. If you are unsure of the answer, analyze the question and the answers. For example:

A rocker panel is a structural member of which vehicle construction type?

A. Front-wheel drive
B. Pickup truck
C. Unibody
D. Full-frame

Analysis:

This question asks for a specific answer. By carefully reading the question, you will find that it asks for a construction type that uses the rocker panel as a structural part of the vehicle.

Answer A is wrong. Front-wheel drive is not a vehicle construction type.

Answer B is wrong. A pickup truck is not a type of vehicle construction.

Answer C is correct. Unibody design creates structural integrity by welding parts together, such as the rocker panels, but does not require exterior cosmetic panels installed for full strength.

Answer D is wrong. Full-frame describes a body-over-frame construction type that relies on the frame assembly for structural integrity.

Therefore, the correct answer is C. If the question was read quickly and the words "construction type" were passed over, answer A may have been selected.

EXCEPT Questions

Another type of question used on ASE tests has answers that are all correct except one. The correct answer for this type of question is the answer that is wrong. The word "EXCEPT" will always be in capital letters. You must identify which of the choices is the wrong answer. If you read quickly through the question, you may overlook what the question is asking and answer the question with the first correct statement. This will make your answer wrong. An example of this type of question and the analysis is as follows:

All of the following are tools for the analysis of structural damage EXCEPT:

A. height gauge.

B. tape measure.

C. dial indicator.

D. tram gauge.

Analysis:

The question really requires you to identify the tool that is not used for analyzing structural damage. All tools given in the choices are used for analyzing structural damage except one. This question presents two basic problems for the test-taker who reads through the question too quickly. It may be possible to read over the word "EXCEPT" in the question or not think about which type of damage analysis would use answer C. In either case, the correct answer may not be selected. To correctly answer this question, you should know what tools are used for the analysis of structural damage. If you cannot immediately recognize the incorrect tool, you should be able to identify it by analyzing the other choices.

Answer A is wrong. A height gauge *may* be used to analyze structural damage.

Answer B is wrong. A tape measure may be used to analyze structural damage.

Answer C is correct. A dial indicator may be used as a damage analysis tool for moving parts, such as wheels, wheel hubs, and axle shafts, but would not be used to measure structural damage.

Answer D is wrong. A tram gauge *is* used to measure structural damage.

Technician A, Technician B Questions

The type of question that is most popularly associated with an ASE test is the "Technician A says. . . Technician B says. . . Who is right?" type. In this type of question, you must identify the correct statement or statements. To answer this type of question correctly, you must carefully read each technician's statement and judge it on its own merit to determine if the statement is true.

Typically, this type of question begins with a statement about some analysis or repair procedure. This is followed by two statements about the cause of the problem, proper inspection, identification, or repair choices. You are asked whether the first statement, the second statement, both statements, or neither statement is correct. Analyzing this type of question is a little easier than the other types because there are only two ideas to consider although there are still four choices for an answer.

Technician A, Technician B questions are really double true or false questions. The best way to analyze this kind of question is to consider each technician's statement separately. Ask yourself, is A true or false? Is B true or false? Then select your answer from the four choices. An important point to remember is that an ASE Technician A, Technician B question will never have Technician A and B directly disagreeing with each other. That is why you must evaluate each statement independently. An example of this type of question and the analysis of it follows.

Structural dimensions are being measured. Technician A says comparing measurements from one side to the other is enough to determine the damage. Technician

B says a tram gauge can be used when a tape measure cannot measure in a straight line from point to point. Who is right?

A. A only

B. B only

C. Both A and B

D. Neither A nor B

Analysis:

With some vehicles built asymmetrically, side-to-side measurements are not always equal. The manufacturer's specifications need to be verified with a dimension chart before reaching any conclusions about the structural damage.

Answer A is wrong. Technician A's statement is wrong. A tram gauge would provide a point-to-point measurement when a part, such as a strut tower or air cleaner, interrupts a direct line between the points.

Answer B is correct. Technician B is correct. A tram gauge can be used when a tape measure cannot be used to measure in a straight line from point to point.

Answer C is wrong. Since Technician A is not correct, C cannot be the correct answer.

Answer D is wrong. Since Technician B is correct, D cannot be the correct answer.

Most-Likely Questions

Most-likely questions are somewhat difficult because only one choice is correct while the other three choices are nearly correct. An example of a most-likely-cause question is as follows:

The most likely cause of reduced turbocharger boost pressure may be a:

A. westgate valve stuck closed.

B. westgate valve stuck open.

C. leaking westgate diaphragm.

D. disconnected westgate linkage.

Analysis:

Answer A is wrong. A westgate valve stuck closed increases turbocharger boost pressure.

Answer B is correct. A westgate valve stuck open decreases turbocharger boost pressure.

Answer C is wrong. A leaking westgate valve diaphragm increases turbocharger boost pressure.

Answer D is wrong. A disconnected westgate valve linkage will increase turbocharger boost pressure.

LEAST-Likely Questions

Notice that in most-likely questions there is no capitalization. This is not so with LEAST-likely type questions. For this type of question, look for the choice that would be the least likely cause of the described situation. Read the entire question carefully before choosing your answer. An example is as follows:

What is the LEAST likely cause of a bent pushrod?

A. Excessive engine speed

B. A sticking valve

C. Excessive valve guide clearance

D. A worn rocker arm stud

Analysis:

Answer A is wrong. Excessive engine speed may cause a bent pushrod.

Answer B is wrong. A sticking valve may cause a bent pushrod.

Answer C is correct. Excessive valve clearance will not generally cause a bent pushrod.

Answer D is wrong. A worn rocker arm stud may cause a bent pushrod.

Summary

There are no four-part multiple-choice ASE questions having "none of the above" or "all of the above" choices. ASE does not use other types of questions, such as fill-in-the-blank, completion, true-false, word-matching, or essay. ASE does not require you to draw diagrams or sketches. If a formula or chart is required to answer a question, it is provided for you. There are no ASE questions that require you to use a pocket calculator.

Testing Time Length

An ASE test session is four hours and fifteen minutes. You may attempt from one to a maximum of four tests in one session. It is recommended, however, that no more than a total of 225 questions be attempted at any test session. This will allow for just over one minute for each question.

Visitors are not permitted at any time. If you wish to leave the test room, for any reason, you must first ask permission. If you finish your test early and wish to leave, you are permitted to do so only during specified dismissal periods.

You should monitor your progress and set an arbitrary limit to how much time you will need for each question. This should be based on the number of questions you are attempting. It is suggested that you wear a watch because some facilities may not have a clock visible to all areas of the room.

4 An Overview of the System

Automobile Parts Specialist (Test P2)

The following section includes the task areas and task lists for this test and a written overview of the topics covered in the test.

The task list describes the actual work you should be able to do as a technician that you will be tested on by the ASE. This is your key to the test and you should review this section carefully. We have based our sample test and additional questions upon these tasks, and the overview section will also support your understanding of the task list. ASE advises that the questions on the test may not equal the number of tasks listed; the task lists tell you what ASE expects you to know how to do and be ready to be tested upon.

At the end of each question in the Sample Test and Additional Test Questions sections, a letter and number will be used as a reference back to this section for additional study. Note the following example: **C.14.10**.

Task List

C. Vehicle Systems Knowledge (36 Questions)
14. Miscellaneous (2 Questions)

Task C.14.10   Recommend proper application and usage of chemicals.

Example:
  119. Materials that become unstable and are likely to burn, explode, or give off toxic fumes are called:
    A. flammable materials.
    B. corrosive materials.
    C. reactive materials.
    D. toxic materials.   (C.14.10)

Analysis:
  Question #119
  Answer A is wrong. Flammable materials are those that easily catch fire or explode.
  Answer B is wrong. Corrosive materials are those that will dissolve metal.
  Answer C is correct. Reactive materials are those that become unstable and are likely to burn, explode, or give off toxic fumes.
  Answer D is wrong. Toxic materials cause injury or death from contact, ingestion, or inhalation.

Task List and Overview

A. General Operations (10 Questions)

Task A.1

Calculate discounts/percentages.

    a. To calculate percent discount:
1. Convert the percent into a decimal number by moving the decimal two places to the left. For example, 10 percent (10%) becomes 0.10.
2. Multiply the decimal number by the original price to get the amount of the discount. For example, 10% of $25.00 is $2.50 ($0.10 \times \$25.00 = \$2.50$).

    b. To calculate percent profit:
1. Subtract the cost from the selling price. For example, if the selling price is $15.00 and the cost is $10.00, the profit is $5.00. ($\$15.00 - \$10.00 = \$5.00$).
2. Divide the profit by the cost; $5.00 divided by $10.00 is 0.5.
3. Convert this number into a percentage by moving the decimal two places to the right; 0.5 or 0.50 becomes 50 percent. The profit is 50 percent.

Task A.2

Calculate special handling charges.

    The customer received $95.00, which is 95% of $100.00. The restock fee, usually expressed as a percentage, is the fee charged for having to handle a returned part.

    a. To calculate the restocking fee, consider 5 percent:
1. Convert the percent into a decimal number by moving the decimal two places to the left; 5 percent becomes 0.05.
2. Multiply the decimal number by the original price to get the amount of the restocking fee; $0.05 \times \$100.00$ is $5.00.
3. Subtract the restocking fee ($5.00) from the original price ($100.00) to get the amount of money to be returned to the customer. $\$100.00 - \$5.00 = \$95.00$.

Task A.3

Identify and convert units of measure.

    If a customer came into your store and needed 5 liters of oil, how many quarts would you give him? One quart is equal to 0.9464 liters, so one quart is almost equal to one liter. Therefore, 5 liters is approximately equal to 5 quarts. In actual practice, this amount will leave the engine ¼ quart low on oil but within the safe operating range.

    There are 61 cubic inches to a liter. To convert to cubic inches multiply the liters by 61 to get cubic inches. For example, a 5 liter engine is the same as a 305 cubic inch engine ($5 \times 61 = 305$). The metric unit of temperature measurement is degrees Celsius (°C). To convert from degrees Fahrenheit (°F) to degrees Celsius using a short formula, first subtract 32 then multiply by 0.555. For example, 212°F is equal to 99.9°C ($212 - 32 = 180$, and $180 \times 0.555 = 99.9$). Using a long formula, which yields a more accurate conversion, 212°F is equal to 100°C.

Task A.4

Determine alphanumeric sequences.

    An alphanumeric listing places a series of numbers and letters in order starting from the left digit and working across to the right. If the numbers are the same, the order sequence continues with the letter A, then B, and so forth.

Task A.5

Determine sizes with precision measuring tools and equipment.

    Take for example the figure reading for question #5 in the Sample Test, 0.184 inch. Because this is a 0–1 inch micrometer, the reading must be between 0 and 1. Since the 1 is the last complete unit visible on the horizontal line, the first measurement is 0.100. Three ¼₀-inch marks are visible after the 1, so the second measurement is $3 \times 0.025 =$

0.075. The horizontal line nearly lines up with the 9 on the vertical line, so 0.009 is the third measurement. Adding up the three measurements (0.100 + 0.075 + 0.009) gives a total reading of 0.184.

Task A.6 — Perform money transactions (cash, checks, and credit cards).

The customer should always pay the amount that is on the total invoice line.

Cash sales are often the most profitable for a business; little paperwork is necessary and, usually, little time is required. With a cash sale, cash flow is not impeded. Offering discounts for cash purchases and making sure cash customers are handled courteously and efficiently encourage customers to pay with cash. The following are a few points for conducting a cash sale effectively:

a. Greet the customer and determine his or her needs. Write this information down to ensure accuracy.
b. Refer to catalogs, note the necessary information, and check inventory.
c. Pull the merchandise and suggest related items and/or services that can benefit the customer.
d. Fill out an invoice and include any applicable discounts.
e. Accept payment for the purchase and make change, if necessary.
f. Thank the customer, remind him or her of the warranty policy, and invite him or her to return.

Task A.7 — Perform sales and credit invoicing.

Charge account sales encourage large purchases or purchases from customers who do not want to carry cash or do not have the cash. Counter personnel must be familiar with the charge policy in their workplace to ensure accuracy and customer satisfaction.

Task A.8 — Interact with management and fellow employees.

Parts specialists interact with people. They work with customers, suppliers, manufacturers, and numerous other segments of the aftermarket industry. The ability to communicate is essential, whether it is explaining the features and benefits of a product to a customer, placing an order with a warehouse distributor, or routing the proper paperwork and data through the various departments in the jobber store.

Interacting with others is not always easy. When personalities clash, misunderstandings will occur. A parts specialist or manager must be a problem solver, an astute negotiator, and a person who gains satisfaction through serving and helping others.

Task A.9 — Understand the value of housekeeping skills (facility, work stations, and back room).

It is the responsibility of all employees to keep all areas of the store clean and orderly. All employees suffer from lost sales if the floors, shelves, and displays are a safety hazard or do not appeal to the customer.

Task A.10 — Assist with new employee and customer training.

The parts specialist should seek whatever help is necessary in order to sell the correct parts and satisfy the customer. In addition, experienced parts specialists are often asked to help train new employees.

Task A.11 — Demonstrate proper safety practices.

Shipping, receiving, and stocking materials requires physical exertion. Knowing the proper way to lift heavy materials is important. Always lift and work within your ability and seek help from others when you are not sure if you can handle the size or weight of the material or object. Auto parts, even small, compact components, can be surprisingly heavy or unbalanced. Always size up the lifting task before beginning.

**Task
A.12**

Identify proper handling of regulated and/or hazardous materials.

The Environmental Protection Agency (EPA), Occupational Safety and Health Administration (OSHA), and other state and local agencies have strict guidelines for handling these materials. OSHA's Hazard Communication Standard (HCS), commonly called the "Right-to-Know Law," applies to all companies that use or store any kind of hazardous chemicals that workers might come in contact with, including: solvents, caustic cleaning compounds, abrasives, cutting oils, and other hazardous materials.

Compliance with the law requires a system for labeling hazardous chemicals and maintaining Material Safety Data Sheets (MSDS). These sheets must be made available to employees, informing them of the dangers inherent in chemicals found in the workplace.

**Task
A.13**

Identify potential security risks.

Expensive items should be displayed behind the counter in locked cases. Parts specialists should give a lot of thought to what gets displayed near entrances and exits. In the moment that a parts specialist is busy researching parts in a catalog, a thief can easily slip something under his or her jacket and glide through the door. It is best to keep the areas around the entrances clear, or to display only large, heavy, or awkward merchandise near the door. Barrels of oil and bags of floor sweep are safe choices for displays near entrances.

**Task
A.14**

Identify parts industry terminology.

An invoice should be completed for each sale in the store. These invoices are used to track inventory and customer purchases for billing purposes. A purchase order is used to allow a company to purchase parts. It will describe the parts and quantity of parts to be purchased along with billing information. A stock order is used by the store to order more stock from the suppliers. A back order refers to merchandise ordered from a supplier but not shipped, due to the supplier being out of stock. An emergency order is an order placed with the supplier on a routine basis, usually weekly or biweekly.

**Task
A.15**

Understand the value of company policies and procedures.

Company policies and procedures define how certain things should be handled. They are also the guidelines for making decisions. The policies will dictate who gets how much of a discount. They will define the return policy and other customer service routines. Procedures define how things need to be reported and recorded, as well as the steps to be taken when something unusual comes up. By adhering to and understanding the company's policies and procedures, a parts specialist will be able to offer consistent customer service.

B. Customer Relations and Sales Skills (11 Questions)

**Task
B.1**

Identify customer types (do-it-yourselfer and professional installer).

Do-it-yourselfers usually need more information than do professional automotive customers. These customers usually need and expect good advice on parts and methods. They view the parts specialist as an expert, and the parts specialist should utilize product knowledge and catalog skills to give the advice needed. Nothing should be sold to a customer that is not actually needed.

One way that a parts specialist can assist do-it-yourselfers without taking too much time away from other customers is to have printed how-to information available. Once the parts specialist determines what work the customer will be doing, the parts specialist can provide something to read while assisting other customers. After the customer reads the information, the parts specialist can clarify some important points and answer questions. Each customer's needs can be met without anyone waiting a long time.

Task B.2

Identify customer needs.

A parts specialist's opening question should be designed to put the customer at ease and elicit as much data as possible. The question, "May I help you?" is the worst possible opening because it invites the opportunity for the customer to reply, "No." Instead, the parts specialist should ask, "How can I help you?" Another good opening is, "Can I show you anything in particular or would you like to browse around first?"

These questions restrict answers to the positive and give the customer a way out without having to say, "No." The customer's response will generally indicate whether he or she has a defined objective, is just looking, or is seriously considering a purchase.

Task B.3

Provide information.

Some customers simply lack confidence rather than skill. Providing information can boost confidence. The parts specialist's greatest impact on customers' confidence is his or her own attitude about their abilities. Avoid speaking down to customers, and converse with them according to their actual abilities. Some problems may be beyond the capabilities of some do-it-yourselfers.

Task B.4

Handle customer complaints and returns.

The parts specialist must not be concerned with placing blame because it is not always evident whether or not there is any blame to place. The customer is not always right, but the parts specialist must be very careful when making corrections. Embarrassed customers will probably spend their money somewhere else.

Often the easiest way for the parts specialist to deal with a difficult situation is to put themselves in the customer's shoes. Most people, including parts counter personnel, are angry when something that was purchased causes a problem. Often the anger is directed toward the person who sold it. The parts specialist should take whatever time is necessary to listen to and try to understand the customer's viewpoint.

Task B.5

Acknowledge the customer.

From the time a customer walks through the door until the time he or she leaves, that person should be acknowledged and responded to favorably. Counter personnel should refrain from personal conversations with one another whenever a customer approaches the counter.

Customers usually do not want to feel like a number or just another face in the crowd. The first step toward making a customer feel noticed is to make eye contact with the customer by looking at him or her and saying something such as, "Good morning. I'll be with you in a moment." Eye contact should also be made when the parts specialist is waiting on the customer by looking up from the parts catalog periodically while talking to him or her.

Task B.6

Demonstrate proper telephone skills.

Often the phone will ring while the parts specialist is serving a customer. If no one else is available to answer the phone, the parts specialist should politely ask the customer to wait and then answer the phone. If it appears that the call will take a while, the parts specialist should explain to the caller that he or she is assisting another customer at the moment. After taking pertinent information, the parts specialist should state that he or she will return the call as soon as possible or that the caller should phone back in a few minutes. That way, neither the caller nor the counter customer is kept waiting very long.

Since a caller cannot see the parts specialist's facial expressions, gestures, or body language, all communications must be accomplished with words, tones, timing, and inflection. Parts counter personnel must remember to speak clearly and slowly enough to be understood and must request that the customer do so if not clearly understood.

**Task
B.7**

Obtain pertinent application information.

Defensive selling implies a method of selling that protects the interests of the store in response to do-it-yourselfers who are not always clear and methodical in their diagnosis of a problem. These customers will sometimes purchase the wrong part, install it, find that the problem has not been solved, and then try to return the part as defective. Parts that are sold and returned in this manner are so done at the expense of the store. For this reason, many stores have a policy that does not allow the return of parts that have been installed.

**Task
B.8**

Present a professional image.

The appearance of the store, as well as the appearance of counter personnel, influences the customer. A clean, neat parts specialist establishes a positive image with customers, and a clean, neat store is equally important.

Although a parts specialist is rarely in charge of decor or layout, he or she can maintain the appearance of the store in several ways:
   a. Keep a rag handy for wiping the counter so that it is free of dirt and grime from used parts.
   b. Keep the counter free of clutter such as small parts, notes, paper clips, and other such items.
   c. Keep the displays organized and well stocked.
   d. Keep the stock and supplies neatly shelved to improve appearance.
   e. Assist with basic clean-up by throwing away empty parts wrappers and labels and by removing empty boxes from sight.

**Task
B.9**

Recognize the value of selling related items.

Selling related parts when a customer buys a particular item can boost profits by 30 percent or more. Selling related parts also makes sense when the parts specialist remembers that a customer's problem might not be solved solely by replacing a faulty part if the related hardware or chemicals are not also up to peak performance. For example, if selling a wheel cylinder, suggest brake fluid.

If the parts specialist suggests related items on the mechanic's first visit, both the customer and the parts specialist can save a lot of time. The customer will be very appreciative of the parts specialist's initial time investment, and with this type of service will more than likely return for parts again.

**Task
B.10**

Identify products' features and benefits.

When the parts specialist carries many different brands of the same product, the difference between them should be explained. If the store carries, for example, four different kinds of brake pads, the parts specialist should identify the material, size, cost, warranty, and name brand differences. The parts specialist should let the customer decide on which part to buy. The parts specialist can make recommendations but the final decision should be the responsibility of the customer.

**Task
B.11**

Handle objections.

After some experience, a parts specialist will know when a customer is ready to buy and when a customer is there to browse. A parts specialist should not push the sale of any parts that a customer may not need. Part of a parts specialist's job is to be able to identify the different customer types and determine their needs.

**Task
B.12**

Balance telephone and in-store customers.

If a parts specialist must put a telephone customer on hold, he or she should identify the name of the store and politely state something such as, "Please hold one moment." The parts specialist should never pick up the phone and press the hold button without saying anything, and he or she should never push the button before finishing a sentence or phrase. If the parts specialist knows that it will take a while before he or she can talk with the caller, the parts specialist should ask that the caller call back rather than be put on hold for a long period of time.

Task B.13
Promote store services and features.

A good parts specialist should always promote sales and store services. If a customer asks for some brake pads, the parts specialist should let the customer know that they have a machine shop and can turn his rotors or drums if needed. The parts specialist can also let the customer know that they have brake fluid and other related brake parts in stock to further promote the store sales.

Task B.14
Promote upgraded products.

Selling related items is extremely important, not only to build profits, but also to help the customer achieve the safest possible results. Displays should be set up to advertise the complete service job. Brakes pads, shoes, and hardware once came in fairly plain, dull boxes, but today, most manufacturers have upgraded their packaging graphics to the point where these products can be used to build attractive, effective displays in the front of the store.

Task B.15
Solve customer problems.

A parts specialist should help the customer get the right part each and every time. If a customer comes in for a starter while carrying a pair of jumper cables, the parts specialist might ask, "What was your car's problem that lead you to believe that it needs a starter?" The customer may say, "It will not start without jump-starting the battery." The parts specialist should then explain that the battery could be the problem. This way the parts specialist will not only get the sale for the battery but will not have sold the customer a part that did not fix the vehicle.

Task B.16
Close sale.

Closing simply means getting a commitment from the customer. Successful salespersons usually share certain characteristics. First, effective closers expect the sale. The very best closers are certain that they will bring each interview to a successful close. Their confidence might be justified, or it might be the product of an inflated ego, but, whatever the source, that expectation of success often results in sales. Second, good closers always let the customer know that they expect the sale. Finally, many parts specialists do a fantastic job of qualifying prospects, matching benefits to needs, laying a foundation of agreement, handling any objections, and then stopping and waiting for customers to start shelling out their money.

C. Vehicle Systems Knowledge (36 Questions)
1. Engine Mechanical Parts (3 Questions)

Task C.1.1
Identify major components.

The engine is a device that burns fuel to produce mechanical power to convert heat energy into mechanical energy. In a passenger car or truck, the engine provides the power to drive the wheels through the transmission—usually by means of a clutch or torque converter and driving axle. Both gasoline and diesel automotive engines are internal combustion engines. The combustion or burning that creates heat energy takes place inside the engine. These systems require an air/fuel mixture that is delivered to the combustion chamber with exact timing. The engine must be constructed to withstand the temperatures and pressures created by fuel burning.

Both gasoline and diesel engines share the same major parts, except the diesel engine does not have spark plugs to ignite the air/fuel mixture. They, instead, have glow plugs.

Task C.1.2

Identify component function.

There are usually two valves at the top of the cylinder. The air/fuel mixture enters the combustion chamber through an intake valve and leaves through an exhaust valve. The valves are accurately machined openings. A valve is said to be seated as closed when it rests in its opening. When the valve is pushed off its seat, it opens.

A rotating camshaft, connected to the crankshaft, opens and closes the intake and exhaust valves. Cams are raised sections of the shaft, or collars, with high spots called lobes. As the camshaft rotates, the lobes rotate and push away a spring-loaded valve tappet. The tappet transfers the motions to a pushrod and perhaps a rocker arm to open the valve by lifting it off its seat.

Task C.1.3

Identify related items.

Inspect the flywheel for scoring and cracks in the clutch contact area. Minor score marks and ridges may be removed by resurfacing the flywheel. Mount a dial indicator on the engine flywheel housing, and position the dial indicator stem against the clutch contact area on the flywheel. Rotate the flywheel to measure the flywheel runout. If the flywheel runout exceeds specifications it must be replaced.

Valve spring retainers and locks must be checked for wear, scoring, or other damage. If any of these conditions are noted, replace the components. The valve lock grooves on the valve stems must be inspected for wear, particularly round shoulders. If found to be uneven or rounded, replace the valve.

Task C.1.4

Provide use and installation instructions.

The causes of excessive oil consumption, such as worn rings, valve guide seals and stems, turbocharger seals, scored cylinder walls, or oil leaks, must be understood. Unusual exhaust gas colors and their causes are:

a. Blue exhaust: excessive oil consumption; may be more noticeable on acceleration and deceleration.
b. Black exhaust: rich air/fuel ratio, excessive fuel consumption.
c. Gray exhaust: coolant leaking into the combustion chambers; may be more noticeable when the engine is first started.

Normal exhaust noise contains steady pulses at the tailpipe. A puff noise in the exhaust at regular intervals usually indicates a cylinder misfire caused by a compression, ignition, or fuel system defect. Erratic exhaust pulses at the tailpipe indicate a rough idle condition caused by ignition or fuel system defects. Excessive exhaust noise indicates a leak in the exhaust system.

A high-pitched squealing noise during hard acceleration may be caused by a small leak in the exhaust system, particularly in the exhaust manifolds or exhaust pipe. An intake manifold vacuum leak causes a high-pitched whistle at idle and low speeds that then gradually decreases when the engine is accelerated and the intake vacuum decreases.

An excessive sulfur smell in the exhaust indicates a rich air/fuel ratio on vehicles equipped with a catalytic converter.

If the cylinder is working normally, a noticeable decrease in engine speed occurs when the cylinder misfires. If there is very little decrease when the analyzer causes a cylinder to misfire, the cylinder is not contributing to engine power. Under this condition the engine compression, ignition system, and fuel system should be checked to locate the cause of the problem. An intake manifold vacuum leak may cause a cylinder misfire with the engine idling or operating at low speed. If this problem exists, the misfire will disappear at a higher speed when the manifold vacuum decreases. When all cylinders provide the specified drop in engine speed, all are contributing equally to the engine power.

2. Cooling Systems (2 Questions)

Task C.2.1

Identify major components.

The burning of the air/fuel mixture in the combustion chambers of the engine produces high amounts of heat. If left unchecked, this heat could damage and warp metal engine parts. To avoid this, engines are equipped with a cooling system. The cooling system is a system of parts that circulates coolant through the engine to remove heat.

The most common method used to cool an engine is to circulate a liquid coolant through the engine block and cylinder head. Liquid cooling is preferable to air cooling because it is less noisy and is better able to maintain a constant temperature at the cylinders. It also lets the engine operate more efficiently and makes a ready supply of hot coolant available to operate a heater for the in-vehicle compartment.

A cooling system is made up of the following components:

a. A pump, which circulates the coolant through the system. The coolant is a mixture of water and antifreeze.

b. Water jackets, which are cored passages in the cylinder block and cylinder head that carry the coolant around the cylinder's combustion chambers.

c. A radiator, which is used to transfer the coolant's heat to the outside (ambient) air as the coolant flows through its tubes.

d. A fan, engine or electrically driven, which pulls cool outside air through the fins of the radiator to help cool the coolant.

e. A radiator pressure cap, to maintain pressure in the system to raise the boiling point of the coolant. Also to provide relief from excess pressure or vacuum.

f. Hoses, which are used to connect the components of the system to each other.

g. A thermostat, which is used to block off circulation in the cooling system until a preset temperature is reached. This is an aid to speed engine warm-up and to keep the engine temperature to a predetermined level.

h. A dash-mounted temperature indicator, to alert the operator in case of overheating.

Task C.2.2

Identify component function.

The water pump is belt driven by the engine crankshaft. The engine cooling fan can be mounted on the water pump or operated electrically. Coolant is pumped or circulated by the water pump from the bottom or side of the engine block and heads. The internal passages are often called the water jackets. The water pump then pumps the coolant through the thermostat to the top of the radiator.

There are two basic types of radiator designs. They are distinguished from one another by the direction of the coolant flow and location of the two tanks. In the downflow radiator, the coolant flows from the tanks downward to the bottom tank. In a crossflow type, the tanks are located at either side, and the coolant flows across the radiator core from tank to tank.

Cool coolant flows from the radiator outlet tank to the engine, and hot coolant is returned to the inlet tank. Turbulence created by the water pump creates minute bubbles in the coolant, which can act to reduce its efficiency. This and the fact that hot coolant expands necessitates an air, or expansion, space for the bubbles to dissipate and the coolant to expand into, thereby allowing the radiator to contain only coolant, not air.

Task C.2.3

Identify related items.

Inspect the oil pump pressure relief valve for sticking and wear. If this valve sticks in the closed position, oil pressure is excessive. A pressure relief valve stuck in the open position results in low oil pressure.

Measure the thickness of the inner and outer rotors with a micrometer. When this thickness is less than specified on either rotor replace the rotors or the oil pump. The following oil pump measurements should be performed with a feeler gauge: (1) measure pump cover flatness with a feeler gauge positioned between a straightedge and the cover, (2) measure the clearance between the outer rotor and the housing, (3) measure the clearance between the inner and outer rotors with the rotors installed, and (4) measure the clearance between the top of the rotors and a straightedge positioned across the top of the oil.

Cooling system hoses should be inspected periodically for swelling, hardening, chafing, leaks, and collapsing. If any of these conditions are noted, replacement is necessary. Hose clamps should be inspected to make sure they are tight. Some radiator hoses contain a wire coil inside them to prevent hose collapse as the coolant temperature decreases. Be sure to include heater hoses and bypass hoses in the inspection. Prior to replacing a hose, the coolant must be removed from the radiator.

Task C.2.4

Provide use and installation instructions.

With the engine not running, grasp the fan blades, or the water pump hub, and try to move the shaft from side to side to check for looseness in the water pump bearing. If any side-to-side movement is noted in the bearing, water pump replacement is required.

Check for coolant leaks, rust, or residue at the water pump drain hole in the bottom of the pump and at the inlet hose connected to the pump. When coolant is dripping from the pump drain hole, replace the pump. The water pump may be tested with a pressure tester connected to the radiator filler neck.

A thermostat that is stuck open reduces coolant temperature and does not cause high coolant level in the recovery tank. The radiator cap should be inspected for a damaged sealing gasket, or vacuum valve. If the pressure cap is damaged, the engine will overheat and coolant will be lost to the coolant recovery system. Under this condition the coolant recovery tank becomes overfilled. If the cap valve is sticking, a vacuum may occur in the cooling system after the engine is shut off and the coolant temperature decreases. This vacuum may cause collapsed cooling system hoses. A pressure tester may be used to test the pressure cap and the entire cooling system. The coolant level should be at the appropriate mark on the recovery container, depending on engine temperature.

3. Fuel Systems (3 Questions)

Task C.3.1

Identify major components.

The fuel system is the system that delivers fuel to the engine cylinders. It consists of a fuel tank and lines, gauge, fuel pump, carburetor or injectors, and intake manifold or fuel rail.

The fuel system supplies a combustible mixture of gasoline and air to the engine's cylinders. To do this, the fuel system must store the fuel, deliver fuel to the metering devices, atomize and mix the fuel with air, and vary the proportion of fuel-to-air to satisfy the many load requirements of the engine.

The fuel system uses several components to accomplish these tasks:

    a. A fuel tank, to store the gasoline in liquid form.

    b. Fuel lines, to carry the liquid from the tank to the other parts of the system.

    c. A pump, to move the liquid from the tank.

    d. A filter, to remove dirt or other harmful particles that might be in the fuel.

    e. A pressure regulator, to keep the fuel pressure below a specified level.

    f. Fuel injectors or a carburetor, to mix the liquid gasoline with air for delivery to the cylinders after the air has passed through an air filter.

    g. An intake manifold, to direct the air/fuel mixture to each of the cylinders.

Task C.3.2

Identify component function.

The fuel system supplies a combustible mixture of gasoline and air to the engine cylinders. In order to do this, it must store the fuel and deliver it to the fuel metering and atomization system, where it is mixed with air to provide the combustible mixture that is delivered in a manner that meets the varying load requirements of the engine.

Task C.3.3

Identify related items.

The intake manifold should be inspected for cracks and pitting. The surface that fits against the cylinder head should be checked for warpage with a straightedge and a feeler gauge. The idle mixture should be adjusted on carbureted engines with the engine at normal operating temperature and idling at the specified speed. After the idle mixture adjustment is completed, the vehicle must meet emission standards.

Excessive blue smoke in the exhaust gas may indicate worn turbocharger seals. The parts specialist must remember that worn valve guide seals or piston rings also cause oil consumption and blue smoke in the exhaust. Check for intake system leaks. If there is a leak in the intake system before the compressor housing, dirt may enter the turbocharger and damage the compressor, or turbine, wheel blades. When a leak is present in the intake system between the compressor wheel housing and the cylinders, turbocharger pressure is reduced. Turbocharger boost pressure may be tested with a pressure gauge connected to the intake manifold. The boost pressure should be tested during hard acceleration.

Task C.3.4

Provide use and installation instructions.

If the positive crankcase ventilation (PCV) valve is stuck in the open position, excessive airflow through the valve causes a lean air/fuel ratio and possible rough idle operation or engine stalling. When the PCV valve or hose is restricted, excessive crankcase pressure forces blowby gases through the clean air hose and filter into the air cleaner. Worn rings or cylinders cause excessive blowby gases and increased crankcase pressure, which force blowby gases through the clean air hose and filter into the air cleaner. A restricted PCV valve or hose may result in the accumulation of moisture and sludge in the engine and engine oil.

Some vehicle manufacturers recommend removing the PCV valve from the rocker arm cover, and placing a finger over the valve with the engine idling. When there is no vacuum at the PCV valve, the valve, hose, or manifold inlet are restricted. Replace the restricted component or components.

Remove the PCV valve from the rocker arm cover and the hose. Shake the valve near your ear and listen for a rattle inside the valve housing. If no rattle is heard, PCV valve replacement is required.

Mechanical or electric fuel pumps should be tested for pressure and flow or volume. A pressure gauge is connected in series in the fuel inlet line on the carburetor or throttle body assembly on throttle body injection (TBI) engines. On most port fuel injection engines the pressure gauge is connected to the Schrader valve on the fuel rail.

4. Ignition Systems (3 Questions)

Task C.4.1

Identify major components.

The ignition system is a major component of a gasoline engine powered vehicle. It consists of the battery, ignition coil, low voltage (primary) wiring, ignition start/run switch, distributor (some vehicles are equipped with a DIS, or distributorless ignition system), high-tension (secondary) wiring, and spark plugs. The ignition system provides the right spark at the right time to the right cylinder to ignite the air/fuel mixture in the combustion chamber.

Task C.4.2

Identify component function.

The computer, ignition module, and position sensor combine to control spark timing and advance. The computer collects and processes information to determine the ideal amount of spark advance for the particular operating conditions. The ignition module uses crank/cam sensor data to control the timing of the primary circuit in the coils. Remember that there is more than one coil in a distributorless ignition system. The ignition module synchronizes the coil's firing sequence in relation to the crankshaft position and firing order of the engine. Therefore, the ignition module takes the place of the distributor.

The ignition module also adjusts spark timing below 400 rpm and when the vehicle's control computer bypass circuit becomes open or grounded. Depending on the type of electronic ignition system, the coils can be serviced as a complete unit or separately. The coil assembly, typically called a coil pack, consists of two or more individual coils.

Task C.4.3

Identify related items.

The mechanical advance controls the spark advance in relation to engine speed. Pivoted weights move outward, by centrifugal force, as the engine speed is increased. This action rotates the cam or reluctor ahead of the distributor shaft. A vacuum advance controls spark advance in relation to the engine load. This advance rotates the pickup or breaker plate in the opposite direction of distributor shaft rotation to provide spark advance.

The coil secondary winding should be tested for shorts, grounds, and opens with an ohmmeter. Always inspect the coil tower for cracks. Secondary coil voltage should be tested with a test spark plug or an oscilloscope.

Task C.4.4

Provide use and installation instructions.

When diagnosing a no-start condition, a 12V test light may be connected from the negative primary coil terminal to ground. If the test light flashes while cranking the engine, the primary circuit is being triggered off and on by the module. A problem in the secondary ignition circuit may be preventing the spark plug from firing. If a test spark plug, connected from each spark plug wire to ground, does not fire when cranking the engine, check the coil, distributor cap and rotor, spark plugs, and wires. If the test light does not flash while cranking the engine, a defective module or pickup coil is indicated.

An ohmmeter should be used to test the secondary coil winding for shorts, opens, and grounds. The secondary coil voltage may be tested with a test spark plug or an oscilloscope. The resistance of the spark plug wires and secondary coil wire may be tested with an ohmmeter.

5. Exhaust Systems (2 Questions)

Task C.5.1

Identify major components.

The exhaust system is a major system in a vehicle. The exhaust system is designed to channel toxic exhaust fumes away from the passenger compartment, to quiet the sound of the exhaust pulses, and to burn, or catalyze, emissions in the exhaust. A typical exhaust system has the following components: exhaust manifold and gasket; exhaust pipe, seal, and connector pipe; intermediate pipe(s); catalytic converter; resonator and/or muffler; tailpipe; and hardware items including heat shields, clamps, gaskets, and hangers.

On some engines, the exhaust system performs several other functions. Some vehicles have a stove on the exhaust manifold that is used to heat the incoming air. A hot-air pipe or hose leads up from the manifold to the air cleaner snorkel and carries heated air to mix with the normal intake air.

Some engines use manifold heat to warm a section of the intake manifold. An exhaust crossover passage directs exhaust under the intake manifold, warming the air/fuel mixture.

Task
C.5.2

Identify component function.

In all types of engines, the exhaust gas flows from the manifold to the exhaust or inlet pipe. The inlet pipe then connects to a series of exhaust pipes comprising a catalytic converter, muffler, and interconnection pipes.

The muffler helps quite the sound of combustion. Some large engine exhaust systems also have a second muffler or resonator. This is used to further muffle the noise of the exhaust. The muffler is simply a can or container with an inlet and outlet pipe attached. Inside the container is a series of baffles designed to deflect the exhaust within the container. The baffling or deflection of the exhaust quiets the sound of combustion as the gases leave the tailpipe.

Task
C.5.3

Identify related items.

Leaks in gaskets, such as rocker arm cover or crankcase gaskets, will result in oil leaks and the escape of blowby gases to the atmosphere. The positive crankcase ventilation (PCV) system also draws unfiltered air through these leaks into the engine. This may result in early wear of engine components when the vehicle is being operated in dusty conditions.

When diagnosing a PCV system, the first step is to check all the engine gaskets for signs of oil leaks. Be sure the oil filler cap fits and seals properly. Check the clean air hose and the PCV hose for cracks, deterioration, loose connections, and restrictions. Check the PCV clean air filter for contamination and replace if necessary. If there is evidence of oil in the air cleaner, check the PCV valve and hose for restriction.

The back pressure reading is taken with a warm engine at 2,500 rpm. A restriction is indicated if the pressure exceeds 1.75 psi (12 kPa). The temperature test is taken after the engine has been running at 2,500 rpm for three minutes. If the carbon monoxide (CO) exceeds 0.3 percent, the oxygen (O_2) exceeds 0.4 percent, and the hydrocarbon (HC) content falls between 120 and 400 parts per million (ppm), the catalytic converter is probably defective.

Task
C.5.4

Provide use and installation instructions.

Many original mufflers and catalytic converters are integral with the interconnecting pipes. When these components are replaced, they must be cut from the exhaust system with a cutting tool. The inlet and outlet pipes on the replacement muffler or converter must have a 1.5 inch (38 mm) overlap on the connecting pipes. Before cutting the pipes to remove the muffler or converter, measure the length of the new component and always cut these pipes to provide the required overlap.

Various vehicle manufacturers recommend different positive crankcase ventilation (PCV) valve checking procedures. Always follow the procedure in the vehicle manufacturer's service manual. Some vehicle manufacturers recommend removing the PCV valve from the rocker arm cover and the hose. Connect a length of hose to the inlet side of the PCV valve and blow air through the valve with your mouth while holding your finger near the valve outlet. Air should pass freely through the valve. If air does not pass freely through the valve, replace the valve.

Connect a length of hose to the outlet side of the PCV valve and try to blow back through the valve. It should be difficult to blow air through the PCV valve in this direction. When air passes easily through the valve, replace the valve.

6. Emissions Control Systems (4 Questions)

Task
C.6.1

Identify major components.

Emissions control systems are required, by federal law, to ensure that vehicles emit low levels of pollutants such as hydrocarbons (HC), carbon monoxide (CO), and oxides of nitrogen (NO$_x$). The Environmental Protection Agency (EPA) establishes emissions standards that limit the amount of these pollutants that a vehicle can emit.

The most common emission control devices and systems are:

a. Positive crankcase ventilation (PCV) system. A system that draws fuel and oil vapors from the crankcase and introduces them into the intake airstream to be burned.

b. Evaporative emission controls (EEC) system. A system that draws fuel vapors from the fuel tank and the carburetor bowl and introduces them into the intake airstream to be burned.

c. Exhaust gas recirculation (EGR) system. A system that introduces exhaust gases into the intake air to reduce the formation of oxides of nitrogen in the combustion chamber.

d. Catalytic converter. A device that burns hydrocarbons and carbon monoxide in the exhaust stream. It also breaks down oxides of nitrogen into harmless nitrogen and oxygen.

e. Air injection system. A system that introduces fresh air into the exhaust stream to cause a second burning of the hydrocarbons in the exhaust.

f. Early fuel evaporation (EFE) system. A system to warm the intake air to reduce the condensation of fuel on the walls of the intake manifold and to speed the warm-up of the vehicle.

Task C.6.2 — Identify component function.

At the heart of the emissions control system is the exhaust gas recirculation (EGR) valve, which opens to allow engine vacuum to siphon exhaust into the intake manifold. The EGR valve contains a poppet valve which lifts off its seat when a vacuum is applied to the diaphragm. Intake vacuum then siphons exhaust into the engine. Like a positive crankcase ventilation (PCV) valve, the EGR valve is a kind of calibrated vacuum leak. The EGR valve uses ported vacuum as its primary vacuum source. Ported vacuum is used because EGR is not needed at idle.

On many late-model vehicles, a positive back pressure or negative back pressure EGR valve is often used. This type of valve requires a certain level of back pressure in the exhaust before it will open when vacuum is applied.

The vacuum control plumbing to the EGR valve usually includes a temperature vacuum switch (TVS) or solenoid to block or bleed vacuum until the engine warms up. On some late-model vehicles with computerized engine controls, the computer actuates the solenoid to further modify the opening of the EGR valve.

Task C.6.3 — Identify related items.

When the engine is not running, or during a backfire, the positive crankcase ventilation (PCV) valve plunger is seated in the housing, and the passage through the valve is closed. If the engine is idling, the high intake manifold vacuum holds the tapered valve plunger nearly closed. When the throttle is opened and the manifold vacuum decreases, the tapered plunger gradually moves downward to provide more PCV valve opening.

If the PCV valve is stuck closed, the air/fuel ratio is richer and hydrocarbon (HC) and carbon monoxide (CO) emissions are higher. A PCV valve stuck in the open position may cause a leaner air/fuel ratio and increased engine idle speed.

If the inside of the air cleaner is contaminated with engine oil, the engine may have excessive blowby, or the PCV valve and connecting hose may be restricted. When the PCV clean air filter in the air cleaner is plugged, higher vacuum is built up in the engine. This may cause damage to the engine gaskets and pull in dirt particles that could cause damage to the engine.

Task C.6.4 — Provide use and installation instructions.

If the airflow is always directed downstream when the bypass mode is completed, the secondary air injection diverter (AIR Diverter) valve or control circuit is not operating properly. If there is no vacuum supplied to the AIR Diverter valve, check the vacuum hoses, AIR Diverter solenoid, and connecting wires. In the upstream mode, both solenoids should be energized, and each solenoid should have 12V at one terminal and a very low voltage at the other terminal. When the voltage is high at the powertrain control module (PCM) side of the AIR Diverter solenoid, the wire from this solenoid to the PCM is open or the PCM is not providing a ground for the wire.

If the catalytic converter rattles when tapped with a soft hammer, the internal components are loose. When this condition is present, the converter should be replaced.

A digital pyrometer may be used to check the catalytic converter. If the converter is operating properly, the converter outlet temperature should be 100°F (38°C) hotter than the inlet temperature.

7. Manual Transmission/Transaxle (2 Questions)

Task C.7.1

Identify major components.

A manual transmission is a manually shifted gearing device in the powertrain that allows variation on the relationship between engine speed and road speed. It is a gearing device that provides variable ratios between the engine output and the differential input.

A manual, or standard, transmission houses a number of individual gearsets that can produce different gear ratios. These gear ratios provide the driver with a selection of speed- and power- (torque-) producing ranges. A typical manual transmission has either three, four, or five forward gear ranges, neutral, and reverse. The driver selects the desired range by disengaging the clutch and moving a shift lever to the required position. This action slides the required gearsets into mesh so the desired output shaft rotation speed is produced.

Transaxles, used on front-wheel-drive vehicles, combine the transmission gearing and the differential gearing into a single unit. The gearsets in the transaxle generate the required gear ratios and pass the power flow on to the differential. The differential gearing in the transaxle creates the final gear reduction and splits the power flow between the left and right front drive axles. Driving axles on front-wheel-drive cars protrude from the sides of the transaxle. These axles can be hollow or solid. The outer end of the axles are fitted to the hubs of the drive wheels. Constant velocity joints mounted on each end of the drive axles allow for steering and suspension (up-and-down) motion without affecting the power flow to the wheels.

Task C.7.2

Identify component function.

When the clutch is engaged, the spring pressure plate forces the disc hard against the face of the flywheel which, when the engine is running, causes the flywheel, disc, and pressure plate to rotate together. When the clutch is disengaged, the pressure plate is pulled away from the disc, allowing the flywheel and pressure plate to turn without rotating the disc. The pressure plate is operated by a clutch release bearing, which is controlled by a series of linkages attached to the clutch pedal. After placing the transmission in gear, the driver gradually engages, or releases, the clutch pedal. The flywheel then transfers its rotary energy to the disc which, in turn, is connected to the transmission by a steel shaft, and the vehicle is set into motion. When the car is moving, each time gears are shifted, the engine's driving force is disconnected from the transmission by disengaging the clutch; if it were not, the transmission gears would be destroyed.

Task C.7.3

Identify related items.

Most external shift linkages and cables require adjusting. A typical adjustment procedure is as follows: Raise the vehicle on a lift and place the shift lever in neutral to begin the shift linkage adjustment. For a lever-type shift linkage, install a ¼-inch (6.35-mm) rod in the adjustment hole in the shifter assembly. Adjust the shift linkages by loosening the rod retaining locknuts and moving the levers until the ¼-inch (6.35-mm) rod fully enters the alignment holes. Tighten the locknuts and check the shift operation in all gears.

Rod-type shift linkages are adjusted with basically the same procedure as lever-type linkages. When the alignment pin is in place, adjust the shift rod so the pin slides freely in and out of the alignment hole.

Fluid leaks in a manual transmission may occur at the extension housing seal, shift cover gasket, vent, backup light switch, drain plug, speedometer drive, or front bearing retainer gasket. If the extension housing seal is leaking, it may be replaced without removing the transmission from the vehicle. When this seal is replaced, always check the fit of the front driveshaft yoke in the extension housing bushing and inspect the yoke for scoring in the seal contact area. A worn extension housing bushing or a scored front driveshaft yoke will cause repeated extension housing seal failure.

Task C.7.4

Provide use and installation instructions.

Worn input shaft splines may cause the clutch disc to stick on these splines resulting in improper clutch release. The gears on the input shaft should be inspected for cracks and pitted, worn, or broken teeth. The bore in each gear and the matching surface on the input shaft should be inspected for roughness, pits, and scoring. Inspect the needle bearings mounted between the gears and the input shaft for roughness and looseness. Use a feeler gauge to measure the clearance between the dog teeth on the third, fourth, and fifth speed gear and the matching blocking ring.

The inspection performed on the input shaft and gears should be repeated on the output shaft and gears. In some transaxles the output shaft assembly is serviced as a complete unit. If any gear or component on the output shaft is worn, the complete assembly must be replaced. A feeler gauge is used to measure clearance between the dog teeth on the first and second speed gear and the matching blocking ring. Inspect the output shaft bearings in the transaxle case for roughness or looseness.

8. Automatic Transmission/Transaxle (2 Questions)

Task C.8.1

Identify major components.

An automatic transmission is a transmission in which gear ratios are changed automatically. It is a gearing device of a vehicle that provides variable ratios between the engine output and the differential input. A combination transmission and axle assembly, common in front-engine, front-drive vehicles is called a transaxle.

An automatic transmission eliminates the use of a mechanical clutch and shift lever. In place of a clutch, it uses a fluid coupling called a torque converter to transfer power from the engine flywheel to the transmission input shaft. The torque converter allows for smooth transfer of power at all engine speeds. Shifting in an automatic transmission is controlled by either a hydraulic system or an electronic system. In hydraulic systems, an intricate network of valves and other components uses hydraulic pressure to control the operation of planetary gearsets. These gearsets generate the three or four forward speeds, neutral, park, and reverse gears normally found on automatic transmissions. Newer electronic shifting systems use electric solenoids to control shifting mechanisms. Electronic shifting is precise and can be varied to suit certain operating conditions.

A transaxle combines the transmission gearing and the differential gearing in a single unit. The transaxle is commonly found on front-wheel-drive vehicles. The gearsets in the transaxle provide the gear ratios and transmit power flow to the differential. The differential gearing creates the final gear reduction and splits power flow between the left and right drive axles.

Constant velocity joints mounted on each end of the drive axles allow for steering and suspension (up-and-down) motion without affecting the power flow to the wheels.

Task C.8.2

Identify component function.

The torque converter provides the fluid coupling. It fits onto the front of the transmission housing and bolts into the engine. Inside its housing are three sets of blades—known as the pump, stator, and turbine—that are immersed in light oil. Energy passes from the engine through the fluid and the blades and is transmitted as mechanical torque to the drive wheels, providing maximum output with little slippage. At idle, the torque converter allows maximum slippage and very little torque output.

The pump supplies fluid under pressure to various piston-in-cylinder servos. These servo units perform the hydromechanical functions of operating the clutches and brakes, which affect gear changes.

Actual gear changes are possible through two sets of planetary gears. These gears are so arranged that those not held by a brake will move. A transmission band is simply a circular strip of spring steel that fits around a drum or shaft, which controls a planetary gear or one of its members. When the servo unit piston closes the band properly around a drum or shaft to keep it from turning, a gear change occurs.

The transmission bands, however, sometimes loosen or get out of adjustment and cannot clamp around the component tightly enough; slipping or erratic shifts can occur, requiring a band adjustment. On most late-model cars, however, band adjustments are seldom necessary after the initial warranty work.

Task C.8.3

Identify related items.

When diagnosing noise problems, pay close attention to the exact conditions when the noise occurs. For example, if a noise occurs with the engine running and the vehicle not moving, the cause is likely to be in the engine or torque converter. When the noise changes with a change in engine speed, the problem may be in the torque converter. If the noise changes with a change in vehicle speed, the problem probably is in the drive line or transmission output shaft.

Automatic transmission fluid (ATF) is pink or red. If the fluid is dark brown or blackish and has a burned odor, the fluid has been overheated, possibly from burned clutches or bands. A milky-colored fluid is most likely caused by coolant contamination from a leaking transmission cooler. If silvery metal particles are found in the fluid, it is an indication of damaged transmission components. If the dipstick feels sticky, and is difficult to wipe clean, the fluid contains varnish, an indication that ATF and filter changes have been neglected.

Task C.8.4

Provide use and installation instructions.

Most automatic transmissions or transaxles have a cable or rod-type shift linkage, connected from the gear selector lever to the transmission lever. If the shift linkage adjustment is not correct, the manual valve is improperly positioned. Improper manual valve position may result in low fluid pressure, improper clutch application, and excessive clutch wear. To check the shift linkage adjustment, remove the shift linkage or cable from the transmission lever. Place the gear selector in the specified position, which often is the park position. Move the transmission lever to the park position and install and tighten the linkage on the transmission lever. Move the gear selector through all the gear positions and be sure there is a detent in each position.

Many band adjustments may be performed externally. In some transmissions, however, the oil pan must be removed to make this adjustment. Improper band adjustment may cause harsh shifting, slipping, band burning, and premature failure.

To complete the band adjustment, loosen the adjuster locknut and tighten the adjuster to the specified torque to seat the band. Loosen the band adjuster the specified number of turns and hold the adjuster in this position while tightening the locknut to the specified torque.

9. Drive Train Components (Includes Driveshafts, Half Shafts, U-Joints, CV Joints, and Four-Wheel-Drive Systems) (2 Questions)

Task C.9.1

Identify major components.

The drive train is made up of all of the components that are required to transfer power from the engine to the driving wheels of the vehicle. The exact components used in a vehicle's drive train depend on whether the vehicle is equipped with rear-wheel drive, front-wheel drive, or four-wheel drive.

Power flow through the drive train of a rear-wheel drive vehicle passes through the clutch or torque converter, through the manual or automatic transmission, and on to the front wheels and a rear differential that drives the rear wheels. This transfer case gearing can be contained in a housing bolted directly to the transmission/transaxle, or it can be located at some point in the driveline running to the rear differential of the vehicle.

Task C.9.2

Identify component function.

The driveshaft is nothing more than an extension of the transmission output shaft. That is, the driveshaft, which is usually made from seamless steel tubing, transfers engine torque from the transmission to the rear driving axle on a rear-wheel-drive vehicle. The yokes, which are either welded or pressed on the shaft, provide a means of connecting two or more shafts together. At the present time, a limited number of vehicles are equipped with fiber composite-reinforced fiberglass, graphite, and aluminum driveshafts. The advantages of the fiber composite driveshaft, other than weight reduction and torsional strength, are fatigue resistance, easier and better balance, and reduced interference from shock loading and torsional problems. The U-joint allows two rotating shafts to operate at a slight angle to each other.

The constant velocity (CV) joint is the driveline on front-wheel-drive vehicles. Most front-wheel-drive cars have two driveshafts, one for each wheel. Each shaft has a CV joint on each end. CV joints nest to the transmission inboard joints, or plunging joints. Most inboard joints, which ride in a tulip shaped housing, use roller bearings on the end of the shaft. This allows the inboard joint to plunge when the shaft changes in length as the wheel follows the road. The CV joint at the wheel hub is an outboard joint. It is fixed and does not change in length.

Task C.9.3

Identify related items.

Most constant velocity (CV) joints are replaced as a complete assembly. Prior to removing the CV joint boot, mark the inner end of the boot in relation to the drive axle so the boot may be installed in the original position. When reassembling the joint always install all the grease in the joint that is provided in the repair kit.

Before removing the driveshaft, mark it in relation to the differential flange so this shaft may be installed in its original position. After the universal joint (U-joint) retaining clips are removed from the driveshaft, a vise and the proper size socket may be used to remove the spider from the yoke.

In a double Cardan U-joint, the center yoke should be marked in relation to the ball tube yokes before disassembly so these components may be reassembled in their original location. The bearing caps in this type of joint should be removed in the proper sequence using the same procedure as followed for a single U-joint. The centering ball and ball seats must be replaced if they are worn or scored.

Refer to the appropriate troubleshooting charts for noise and vibration problems associated with drive train components. To diagnose the problem, be mindful of when the noise occurs, such as on turns, on acceleration or deceleration, at low or high speed, or when in or out of gear, to name a few.

Task C.9.4

Provide use and installation instructions.

Measure the ring gear runout and differential case side play before removing the differential case and ring gear. Be sure to mark the side bearing caps in relation to the housing before removal. Remove the side bearing caps and lift the case and ring gear from the housing.

Before installing the case and ring gear assembly, the side bearings and bearing caps must be lubricated with manufacturer's recommended differential lubricant. Install the case and ring gear assembly with the bearing caps and shims or adjuster nuts. Be sure the bearing cap threads are properly seated in the adjuster nut threads when the bearing caps are installed. Tighten the bearing cap bolts to the specified torque.

The backlash between the ring gear and pinion gear is adjusted after the differential is assembled. Mount a dial indicator on the differential housing, and position the dial indicator stem against one of the ring gear teeth. Zero the dial indicator and rock the ring gear back and forth against the pinion gear teeth. The ring gear backlash is indicated on the dial indicator. Vehicle manufacturers usually recommend measuring the backlash at several locations around the ring gear.

Side bearing preload limits the amount of differential case movement in the axle housing or carrier. When the differential has threaded adjusters on the outside of the side bearings, loosen the right adjuster and tighten the left adjuster to obtain zero backlash. Turn the right adjuster the specified amount to obtain the proper preload. Then rotate each adjuster the same amount in opposite directions to obtain the specified backlash.

When shims are positioned behind the side bearings to adjust backlash and preload, the differential case is pried to one side and the movement is recorded with a dial indicator or feeler gauge. Service spacers, shims, and feeler gauges are installed on each side of the side bearings to obtain zero side play and zero backlash. Proper shim thickness is calculated to provide the specified backlash and side bearing preload. On some differentials, the proper shims have to be driven into place behind the side bearings with a special tool and a soft hammer.

10. Brakes (3 Questions)

Task C.10.1

Identify major components.

The brake assembly is an assembly of the components of a brake system, including its mechanism for the application of friction forces. It is a system used to stop or slow a vehicle and/or prevent it from moving when stopped or parked. Vehicles are slowed or stopped by activating the brake system. Brakes, located at each wheel, utilize friction to slow and stop the vehicle.

The brakes are activated when the vehicle's operator depresses a brake pedal. The brake pedal is connected to a plunger in a master cylinder, which is filled with hydraulic fluid. The force exerted by the plunger on the hydraulic fluid is transferred through brake hoses and lines to the four brake assemblies.

The two types of brakes used on vehicles are drum brakes and disc brakes. Many use a combination of the two: disc brakes at the front wheels and drum brakes at the rear wheels.

Most vehicles have power-assisted brakes. A brake booster uses manifold vacuum to increase the pressure applied to the plunger in the master cylinder. This lessens the amount of pressure that must be applied to the brake pedal by the operator and increases the responsiveness of the brake system.

Task C.10.2

Identify component function.

The front brakes do as much as 80 percent of the work in stopping a car. Older front/rear split hydraulic systems have given way to dual diagonal split systems. These systems increase stopping power and provide 50 percent of the braking capacity in case of failure in either of the two hydraulic systems.

The actuating system of today's cars employs hydraulic pressure or power-assisted hydraulic pressure. When the driver steps on the brake pedal, the force from the foot creates hydraulic pressure in the master cylinder. This pressure is transmitted through steel lines and hoses to the wheel cylinders/calipers where it turns into mechanical force again to act upon the drums or discs that stop the car.

Task C.10.3

Identify related items.

Brake lines should be inspected for leaks, cracks, rust, kinks, flattened areas, and splits. Damaged brake lines should be replaced rather than repaired. A tube-bending tool should be used to form the necessary brake line bends.

Flexible brake hoses allow for movement between the suspension and the chassis. These hoses should be inspected for cracks, leaks, twists, bulges, loose supports, and internal restrictions. Each time a brake hose is removed, the sealing washer on the male end should be replaced. When a brake hose is installed, always install and tighten the male end first.

Task C.10.4

Provide use and installation instructions.

All brake return springs should be inspected for distortion and stretching. Brake shoes should be cleaned with a shop rag and inspected for broken welds, cracks, wear, and distortion. If the wear pattern on the brake shoes is uneven, the shoes may be distorted. Check all clips and levers for wear and bending. Inspect the brake linings for contamination from oil, grease, or brake fluid. Clean and lubricate adjusting and self-adjusting mechanisms.

The backing plate should be cleaned with an approved brake cleaner that does not allow the release of asbestos dust to the shop air. Inspect the backing plate for distortion, cracks, rust damage, and wear in the brake shoe contact areas. A distorted backing plate may cause brake grabbing. Check the anchor bolt for looseness that may result in brake chatter.

11. Suspension and Steering (3 Questions)

Task C.11.1

Identify major components.

The suspension system comprises the components that support the total vehicle weight. This includes the front and rear suspensions, springs, shock absorbers, torsion bars, axles, MacPherson strut system, and connecting linkages. The steering system is the mechanism that permits the driver to change vehicle direction by turning a wheel inside the vehicle.

The suspension system is designed to support the body and frame, the engine, and the drivelines. Without these systems, the comfort and ease of driving the vehicle would be reduced. Springs and torsion bars are used to support the axles of the vehicle. Two types of springs commonly used are the leaf spring and the coil spring. Torsion bars are made of long spring steel rods. One end of the rod is connected to the frame, while the other end is connected to the movable parts of the axles. As the axles move up and down, the rod twists and acts as a spring. Shock absorbers are used to slow down the upward and downward movement of the vehicle. This action occurs when the car goes over a rough road.

The steering system allows the control of the direction of the vehicle. A steering system includes the steering wheel, steering gear, steering shaft, and steering linkage. The two types of steering gears in use are the rack-and-pinion gear and the recirculating ball gear. The rack-and-pinion gear is the most commonly used. The recirculating ball gear is normally used only on heavy vehicles, such as large pickup trucks, station wagons, and full-size luxury cars. Steering gears provide gear reduction to make turning the wheels easier. On all but some light subcompact and compact models, the steering gear is also power assisted to ease the effort of turning the wheels.

Some vehicles have four-wheel steering. Some use a mechanical gearbox directly linked to the front steering gear to turn the rear wheels, others use a second power steering gearbox that is computer controlled. In low-speed turning mode, the rear wheels turn in the opposite direction of the front wheels to reduce the turning radius. When turning the wheels at high speeds, the rear wheels turn in the same direction as the front wheels so that the back wheels follow the front wheels in quick lane changes and other high-speed maneuvers.

Task C.11.2

Identify component function.

The suspension system keeps the vehicle's tires in solid contact with the road while it lets them move to cushion bumps and jolts, or road shock. The steering system keeps the tires pointed in the right direction and works with the suspension system to allow the wheels to be steered with precise directional control.

The functions of steering and suspension are different. Steering is an operation performed by the driver to direct the vehicle. The suspension is a structural arrangement at each of the four wheels. The suspension system includes the shock absorbers, springs, and control arms, which keep the wheel and axles in their correct positions. The steering system connects the steering wheel to the front wheels to provide the directional control. The steering system is supported by the vehicle frame. The relation of the wheels to the frame determines the type of suspension system.

Task C.11.3

Identify related items.

The shock absorber bounce test is performed when one side of the bumper is pushed downward with considerable weight and then released. A satisfactory shock absorber or strut should complete only one free upward bounce before the vertical chassis movement stops. If the bumper completes more than one free upward bounce, the shock absorber is defective or the shock absorber mountings are loose.

A manual test may be performed on shock absorbers. During this test the lower end of the shock absorber is disconnected, and hand pressure is used to extend and compress the shock. A satisfactory shock absorber should provide a strong and steady resistance to movement on both the compression and rebound strokes.

A loose front wheel bearing may cause steering wander. When two tapered roller bearings are mounted in the front or rear wheel hub, a typical bearing adjustment procedure is as follows: (1) tighten the bearing adjustment nut to 17 to 25 ft.-lbs. (23 to 34 N-m), (2) back off the adjustment nut one-half turn, (3) tighten the bearing adjustment nut to 10 to 15 ft.-lbs. (13.6 to 20.3 N-m), and (4) install the nut retainer and cotter key.

Task C.11.4

Provide use and installation instructions.

On front suspension systems without a caster adjustment, some vehicle manufacturers recommend inspecting suspension components for wear or damage to determine the cause of the improper caster angle. Damaged or worn components such as the lower control arm, upper strut mount, or engine cradle may cause an improper caster angle. The damaged or worn component should be replaced. Some parts manufacturers supply suspension components that provide caster adjustment capabilities. Excessive positive caster may cause harsh riding, rapid steering wheel return, and excessive steering effort. Negative caster causes reduced directional stability while driving straight ahead.

12. Heating and Air Conditioning (3 Questions)

Task C.12.1

Identify major components.

Heating is the use of an apparatus that produces a relatively high degree of warmth under controlled conditions, by natural or mechanical means, to ensure personal warmth and comfort. An air conditioner is a device that is used for the control of temperature, humidity, and movement of air in a given space.

Heating and air conditioning, then, is the process of adjusting and regulating, by heating or refrigerating, the quality, quantity, temperature, humidity, and circulation of air in a space or enclosure.

Task C.12.2

Identify component function.

The compressor is nothing more than a pump driven by the crankshaft via drive belts. It picks up a gaseous refrigerant from the evaporator inside the car and compresses it. The compressor uses an electromagnetic clutch to permit it to be turned off when not needed.

The refrigerant is metered into the evaporator, normally located inside the car on the firewall, at about 30 psi (207 kPa). The refrigerant is in liquid form at this point. Its boiling point at 30 psi (207 kPa) is just above the freezing point of water. The refrigerant therefore tends to boil, absorbing heat from the evaporator.

A blower forces either inside or outside air through the evaporator and into the passenger compartment. As the air passes through the evaporator, heat and moisture are removed. The refrigerant inside the evaporator then passes into the compressor as a gas, where its pressure is increased. Refrigerant pressure as it leaves the compressor is usually 200 psi (1,379 kPa). The refrigerant then enters the condenser, a heat-exchanging coil resembling a radiator and usually located in front of the car's radiator. The high pressure caused by the compressor is put to work at this point and raises the boiling point of the refrigerant to over 150°F (65.6°C). When outside air is passed over the thin tubes and fins of the condenser, it cools and changes the refrigerant back to liquid, losing the heat it picked up from the interior of the car.

The liquefied refrigerant then enters the receiver/dryer, a small tank located next to the condenser or on one of the fender wells. This unit has the job of separating liquid refrigerant from any gas that might have left the condenser, and also filtering the refrigerant through a moisture-absorbing material. Many receiver/dryers include a sight glass that allows the refrigerant to be visually checked for the presence of bubbles. This sight glass can be invaluable in troubleshooting air conditioner problems.

Refrigerant then flows through a liquid line to the expansion valve. This valve is usually located near the evaporator, on or near the fire wall. The valve, shaped like a mushroom on some systems, controls the flow of refrigerant to the evaporator. It provides only the amount of flow that the evaporator can handle.

The heater is a comfort control item used especially in colder climates. It may be part of the air conditioning system or a separate item. The heater core can be considered a miniature version of the radiator. As hot coolant flows through the heater, a fan blows air over the tubes, warming it and delivering it to the passenger compartment. The blower fan, located in the heater housing, forces air through the heater core and into the passenger compartment.

The air conditioning/heating distributor system is a duct system. Outside air enters the system through a grill usually located directly in front of the windshield, and then goes into a plenum chamber from where it can pass through the vehicle's heater core, the air conditioning system evaporator, or into a duct that runs across the fire wall of the car. Outlets in the ducts direct the airflow into the passenger compartment.

Task C.12.3

Identify related items.

The thermal fuse and superheat switch may be tested with an ohmmeter or voltmeter. With the ignition switch and the air conditioner (A/C) switch on, and the ambient switch closed, there should be 12V at terminals B, C, and S on the thermal fuse. If there is 12V at terminal B and 0V at terminals C and S, the thermal fuse is blown. When there is 12V at terminals B and C, but a lower voltage at terminal S, the superheat switch contacts are closed, or the wire from the thermal fuse to the superheat switch is shorted to ground. If there is 12V at terminals B, C, and S on the thermal fuse, but 0V at the compressor clutch, the wire from the thermal fuse to the compressor clutch is open.

When the ohmmeter leads are connected to the compressor clutch terminals, an infinite (∞) reading indicates an open clutch coil. An ohmmeter reading below the specified value indicates a shorted clutch coil.

If there is evidence of oil around the high-pressure relief valve and the system is low on refrigerant, check the system pressures and inspect the condenser for restricted air passages. When the condenser air passages are not restricted and the system pressures are normal, the high-pressure relief valve may be defective.

In some A/C systems, a thermal switch is connected in series with the compressor clutch. This switch usually is mounted in the compressor. Many thermal switches open at 257°F (125°C) and close at 230°F (110°C).

If there is a significant temperature difference between the inlet and outlet pipes, the receiver/dryer is restricted and must be replaced. Frost forming on the receiver/dryer indicates an internal restriction that requires component replacement. Replacement is also necessary if there is moisture in the refrigeration system, often indicated by bubbles or foam in the sight glass or rust contamination in the refrigerant. This indicator, however, should not be confused with a low charge of refrigerant.

Grayish blue particles in the sight glass are an indication that the desiccant in the receiver/dryer has disintegrated and is circulating through the refrigeration system. This condition also requires receiver/dryer replacement. If the refrigerant in the sight glass is red or yellow, leak-detecting dye has been added to the refrigeration system. This condition does not require corrective action. All of the refrigerant must be recovered from the system before receiver/dryer removal.

Task C.12.4

Provide use and installation instructions.

Evaporator replacement is necessary if it is leaking or restricted. Evaporator leaks may be detected with an electronic leak tester with a visual or audible beep. Remove the blower motor resistor assembly from the air conditioning (A/C) heater case. Insert the leak tester probe through this opening to check for evaporator leaks. Refrigerant leaks in the evaporator also are indicated by evidence of oil at the condensate drain hose. Evaporator restriction is indicated by a low side pressure that is considerably lower than specified and inadequate cooling.

Always recover all refrigerant from the system and disconnect the negative battery cable before removing the evaporator. Since the evaporator core usually is mounted in the A/C–heater case, this assembly must be removed from the vehicle to remove the evaporator. Be sure to follow any manufacturer's recommendations relative to air bag safety.

If the heater supplies cold or only slightly warm air continually, the engine thermostat may be stuck open or the temperature door may be stuck in the cold position. A coolant flow control valve stuck in the closed position or restricted also causes cold air discharge from the heater.

Tape a thermometer to the upper radiator hose and allow the engine to idle for fifteen minutes. The thermometer reading should be within a few degrees of the specified thermostat opening temperature. If the thermometer reading is considerably lower than the thermostat rating, the thermostat is defective. Move the A/C–heater controls from MAX cold to the MAX heat position and observe the temperature door linkage. If this linkage does not move, the door is either sticking or the actuator system is otherwise inoperative. When the A/C–heater controls are in the heat position, check the temperature on both sides of the coolant flow control valve. The temperature should be the same on both sides of the valve. If this valve is restricted or closed, the temperature at the inlet will be much higher than at the outlet.

If the A/C–heater system supplies warm or hot air continually, check the refrigeration system pressures. If these pressures are not within specifications, repair the refrigeration system. If the refrigeration pressures are within specification and the air discharge is warm, the temperature blend door may be stuck or binding.

13. Electrical Systems (2 Questions)

Task C.13.1

Identify major components.

The electrical system is any of the systems and subsystems that make up the automobile wiring harnesses, such as the lighting system or starting and charging system.

Vehicles have many circuits that carry electrical current from the battery to the individual components. The total electrical system includes subsystems such as the ignition system, starting system, charging system, lighting system, and accessory systems.

The ignition system provides the energy required for combustion. After the air/fuel mixture is in the combustion chamber and compressed by the piston, it must be ignited. A gasoline engine uses an electrical spark to ignite the mixture. Creating this spark is the role of the ignition system. The electrical power required to create this spark is produced by a coil. The coil transforms the low primary voltage of the battery into a secondary burst of 20,000 volts or more. This surge of high voltage must be timed to arrive at the cylinder at just the right moment during the compression stroke. On many engines, the motion of the piston and the rotation of the crankshaft is monitored by a crankshaft position sensor which electronically tracks the position of the crankshaft and relays that information to an ignition control module. Based on input from the crankshaft position sensor and, in some systems, the electronic engine control computer, the ignition control module then turns the battery current to the coil on and off at just the precise time so that the voltage surge arrives at the cylinder at the right time.

The voltage surge from the coil must be distributed to the correct cylinder since only one cylinder is fired at a time. In some systems, this is the job of the distributor. The distributor is driven by a gear on the crankshaft at one-half the crankshaft speed. It transfers the high-voltage surges from the coil to the high-tension spark plug wires in the correct firing order.

The spark plug wires deliver the current surges to the spark plugs, screwed into the cylinder head. The current jumps across a space between two electrodes on the end of each spark plug and a spark is created, causing the air/fuel mixture in the combustion chamber to ignite.

Many ignition systems do not have a distributor. Instead, these systems have several coils, one for each pair of spark plugs. When a coil is activated by the electronic control module, the voltage surge is sent directly to two spark plugs, which fire simultaneously. One spark plug fires during the compression stroke of a cylinder while the other fires during the exhaust stroke and is wasted. In this way, the electronic control module controls both the timing and the distribution of the coil's spark-producing voltage.

The starting system initiates engine operation. When the ignition key is turned to the START position, a small current flows from the battery to a solenoid or relay switch which, in turn, closes another electrical circuit that allows full battery voltage to the starter motor. The starter motor then turns the flywheel, mounted on the rear of the crankshaft, to start the engine and put all engine parts in motion.

The charging system maintains the battery's state of charge and provides power for all of the electrical systems and accessories. In addition to the battery, the charging system includes the engine-driven alternator (or generator), voltage regulator, dash gauge (or indicator light), and associated wiring.

Task C.13.2

Identify component function.

The starting system operates as follows: When the ignition key is turned to the START position, electric current is sent to the solenoid, and battery voltage is supplied directly to the starter motor. The starter motor then turns a flywheel mounted on the rear of the crankshaft that starts all engine parts in motion. The ignition system provides a spark to the spark plugs which ignite the air/fuel mixture from the carburetor. If all components are in good working condition, the engine should start immediately.

The charging system performs two basic tasks: maintaining the battery's state of charge and providing electrical power for the ignition system, air conditioning/heater, lighting, radio, and all other electrical accessories. Turned by a drive belt that is driven by the engine crankshaft, an alternator converts mechanical energy into electrical energy. When the electrical current flows into the battery, the battery is said to be charging. When the current flows out of the battery, the battery is said to be discharging.

The vehicle must be protected from fire hazards that could occur if a powered circuit is accidentally shorted or grounded. Fuses, circuit breakers, and fusible links can be used to protect circuits.

Task C.13.3

Identify related items.

Continuity in an electric circuit may be tested with a 12V test lamp. First, make certain that the test lamp is not defective by connecting it across the battery. If it does not illuminate, it is defective. To test the circuit, connect one of the test lamp leads to ground. With voltage supplied to the circuit, begin at the battery and connect the test lamp to the various terminals in the circuit. When the test light is not illuminated, the open circuit is between the terminal where the test light is connected and the last terminal where the test light was illuminated.

An ohmmeter has an internal battery power source. Damage may result if this meter is connected to a live circuit. The proper scale on the meter must be selected for the component being tested. For example, when testing a component with a 10,000 Ohm (Ω) resistance, select the ×1000 scale on the meter. The meter would then read 10 ($10 \times 1,000 = 10,000$).

Task C.13.4

Provide use and installation instructions.

Connect the ohmmeter leads to each pair of stator leads to test the stator for an open condition. When the ohmmeter leads are connected from one of the stator leads to the stator frame, the stator is tested for a grounded condition.

The ohmmeter leads may be connected across the rotor slip rings to test the field winding for an open or a shorted condition. Connect the ohmmeter leads from one of the slip rings to the shaft to test the field winding and slip rings for a grounded condition. Connect the ohmmeter leads across each diode, and then reverse the leads to test the diodes. A satisfactory diode provides one low and one high ohmmeter reading, a shorted diode is indicated by two low meter readings, and an open diode provides two high or infinite readings.

Some wiper motors contain a series field coil, a shunt field coil, and a relay. When the wiper switch is turned on, the relay winding is grounded through one set of switch contacts. This action closes the relay contacts, and current is supplied through these contacts to the series field coil and armature. Under this condition the wiper motor starts turning. If the wiper switch is in the high-speed condition, the shunt coil is not grounded and the motor turns at high speed.

When the wiper switch is in the low-speed position, the shunt coil is grounded through the second set of wiper switch contacts. Under this condition current flows through the shunt coil and the wiper switch to ground. Current flow through the shunt coil creates a strong magnetic field that induces more opposing voltage in the armature windings. This opposing voltage in the armature windings reduces current flow through the series coil and armature windings to slow the armature. If the wiper motor fails to park, or parks in the wrong position, the parking switch or cam probably is defective.

Some wiper motors have permanent magnets in place of the field coils. These motors have a low-speed and a high-speed brush. In some of these motors the low-speed brush is directly opposite the common brush, and the high-speed brush is positioned in between these two brushes.

14. Miscellaneous (2 Questions)

Task C.14.1

Identify fastener thread types SAE, VSS, and metric.

Three fastener types commonly used in the automotive trade—the United States Standard (USS), the American National Standard (ANS), and the Society of Automotive Engineers Standard (SAE)—have all been replaced by the Unified National Series. The Unified National Series consists of four basic classifications: (1) Unified National Coarse (UNC or NC), (2) Unified National Fine (UNF or NF) (SAE), (3) Unified National Extra Fine (UNEF or NEF), and (4) Unified National Pipe Thread (UNPT or NFIT).

Task C.14.2

Identify fastener thread diameter and pitch.

The two common metric threads are coarse and fine and can be identified by the letters SI (Systeme International d'Unites or International System of Units) or ISO (International Standards Organization).

To identify the type of threads on a bolt, bolt terminology must be defined. The bolt has several parts. The bolt head is used to torque or tighten the bolt. A socket or wrench fits over the head, which enables the bolts to be tightened. Common U.S. Customary (USC) and metric sizes for bolt heads include sizes given in fractions of an inch (English) and in millimeters (metric).

Task C.14.3

Identify fastener type.

Although some USC and metric head sizes are very close, never use a metric wrench or socket for USC bolts, or vice versa. Tool slippage may cause injury or damage the bolt head.

Bolt diameter is the measurement across the major diameter of the threaded area or across the bolt shank. The thread pitch of a bolt in the English system is determined by the number of threads there are in one inch of threaded bolt length and is expressed in number of threads per inch. The thread pitch in the metric system is determined by the distance in millimeters between two adjacent threads. To check the thread pitch of a bolt or stud, a thread pitch gauge is used. Gauges are available in both English and metric dimensions. Bolt length is the distance measured from the bottom of the head to the tip of the bolt. The bolt's tensile strength, or grade, is the amount of stress or stretch it is able to withstand. The type of bolt material and the diameter of the bolt determine its tensile strength. In the English system, the tensile strength of a bolt is identified by the number of radial lines (grade marks) on the bolt head. More lines mean higher tensile strength. In the metric system, tensile strength of a bolt or stud can be identified by a property class number on the bolt head. The higher the number, the greater the tensile strength.

Task C.14.4

Identify fastener grade.

It is very important to be familiar with the standard bolt indication measurements and grade markings. All bolts in the same connection must be of the same grade. Otherwise, they will not perform equally. Likewise, nuts are graded to match their respective bolts. For example, a grade 8 nut must go with a grade 8 bolt. If a grade 5 nut is used instead, a grade 5 connection will result. The grade 5 nut cannot carry the loads expected of the grade 8 bolt. Grade 8 and critical applications require the use of fully hardened flat washers. They do not dish out like soft wrought washers that cause loss of clamp load.

Bolt heads can pop off because of fillet damage. The fillet is the smooth curve where the shank flows into the bolt head. Scratches in this area introduce stress to the bolt head, causing failure. The bolt head can be protected by removing any burrs around the edges of holes. Also, place flat washers with their rounded, punched side against the bolt head and their sharp side to the work surface.

Task C.14.5

Identify fitting type.

Flare, compression, and pipe are three types of fittings. A flare fitting requires that the end of the tubing be expanded at an angle, or flared. This operation requires the use of a flaring tool. Compression fittings do not require any special tools. A soft metal band, known as a ferrule, is placed between the end of the tubing and the fitting. As the two fittings are tightened together, the ferrule is crimped into a sealing connection between the fittings and tubing. Pipe fittings seal using tapered threads. The more that pipe fittings are tightened, the closer the threads are brought together.

Task C.14.6

Identify fitting size.

Fittings should be manufactured to meet the functional requirements of the Society of Automotive Engineers (SAE), Automotive Service Association (ASA), and the American Society of Mechanical Engineers (ASME) for low, medium, or high pressure line connection service as applicable.

Pipe fittings made of brass or iron can be used with air, water, oil, fuels, and various gases. They may be connected by brass, copper, or iron pipe. These fittings are applied in a wide variety of automotive applications including off-road as well as on-road equipment. Available styles include nipple, tee, elbow, adapter, plug, and cap. Male and female pipe fittings are available in popular sizes from ⅛ inch to 2 inches.

Flare fitting applications include refrigeration equipment, air compressors, oil burners, and most any type of machinery. Flare fittings may be used with thin-wall tubing where application requires joints to withstand high pressures (up to 2,800 psi [19,306 kPa]) using 0.030-inch (0.762-mm) wall copper tubing. They may also be used with flareable copper, brass, aluminum, and welded steel hydraulic tubing. Tubing flared at 37 or 45 degrees (as applicable) forms a sound joint resisting mechanical pull-out. It seals and remains leak-free even when disconnected and reconnected. Flare fittings are available in long and short nuts, plugs, unions, elbows, tees, and adapters. They are available in male, female, or male-to-female configuration. Popular flare fitting sizes are from ¼ inch to ⅝ inch.

Compression fittings eliminate the need to flare, solder, or otherwise prepare tubing before assembly. They are for use where excessive vibration or tube movement is not a problem. There are some applications, such as brake service, where compression fittings are not used. Compression fittings are used with copper, brass, or aluminum and are easily installed. All sealing surfaces are precision machined to ensure reliable and leakproof closures. Compression fittings include sleeve(s) and nut(s). They are available in many configurations including unions, elbows, and tees, in popular sizes of ¼ inch to ⅝ inch.

It should be noted that many variations of fitting combination types are available. These include, but are not limited to, flare to pipe, flare to compression, and compression to pipe. These include male-to-female as well as female-to-male provisions. Sizes given are for English. However, many are available in metric sizes as well.

Task C.14.7

Identify body repair and refinishing materials and supplies.

Automotive refinishing materials and procedures have changed considerably in recent years. There are many brands and types of automotive paints now available. As a result, the auto body technician must learn new technologies to be able to apply the various types of automotive finishes.

The terms *paint* and *finish* are generally used to describe many different finishing materials. They are, however, most commonly used when referring to the topcoat or outside layer (about 0.004 inch [0.102 mm] thick) applied to metal or plastic components.

All paints contain three basic ingredients: binders, solvents, and pigments. Special additives are also used in some paints to alter their properties and characteristics. Binders are the resinous, film-forming ingredients that adhere to the substrate, the surface being painted. Pigments are added to the paint to give it color. The amount and type of pigment will somewhat alter the durability, adhesion, flow, and other characteristics of the paint. A number of additives are used in paint to alter its properties and characteristics. Manufacturer's instructions should always be followed when mixing additives with paint.

Primers are used to help improve paint adhesion on a metal or plastic substrate. The general term *primer,* also called prepcoat, implies that the substrate is being prepared for a final coat.

Solvents, also called vehicles, are added to paint so that it can be applied easily. Solvents thin the paint pigments and binders so that they may be sprayed. Solvents are sometimes called volatiles because they vaporize and evaporate easily and rapidly.

Lacquer topcoats dry from the outside in as their solvents evaporate. Chemical hardeners, or isocyanates, are not used with lacquer finishes. One reason that lacquer finishes are so popular is because they dry rapidly. Lacquer remains more or less soluble, allowing a new coat of acrylic lacquer to bond or unite with an old lacquer finish.

In prior years, original equipment manufacturers (OEMs) used a specially formulated lacquer which was heated (baked) in high-temperature ovens to promote drying and curing. These lacquer-based finishes have now been phased out and have been replaced by various types of enamels, such as the following:

a. Alkyd enamel. A favorite for over forty years, alkyd enamel is the least expensive of the enamels. It usually covers in two coats and is sometimes referred to as a synthetic enamel because the alkyd resin is made from petroleum rather than natural sources.

b. Acrylic enamel. An enamel that consists of a solvent blend of binder and pigment materials. They flow out well to fill small imperfections in the substrate. Acrylic enamels must be applied in a totally dust-free environment and permitted a long time to dry.

c. Polyurethane enamel. An enamel used for automotive refinishing that provides a hard, tile-like, high-gloss finish. It has excellent flow-out and appearance, good adhesion, and flexibility. These finishes have been used on aircraft, off-road equipment, and fleet trucks, as well as automobiles. Polyurethane enamels produce a "wet-look" finish.

d. Acrylic-polyurethane enamel. An enamel that weathers well and provides a higher gloss and greater durability than other polyurethane enamels. A two-part finish, it must be applied immediately after being mixed, which activates it.

e. Waterborne acrylic enamel. An enamel used on some vehicles. This finish is baked at more than 300°F (149°C). After baking, the finish is covered with a polyurethane clearcoat. This process is not practical for repair shops since most do not have baking ovens.

Task C.14.8 — Identify hose and tubing types and applications.

The primary purpose of tubing or hose is to transmit fluid, often under high pressure, from one point to another. The power steering pump, for example, moves power steering fluid, via hoses, from the pump to the steering gearbox and returns the fluid, under low pressure, to the pump reservoir. Hoses also function as additional reservoirs and act as sound and vibration dampers.

Hoses are generally a reinforced synthetic rubber material coupled to metal tubing at connecting points. The pressure side must be able to handle pressures up to 1,500 psi (10,343 kPa). For that reason, wherever there is a metal-to-rubber tubing connection, the connection is crimped. Pressure hoses are also subject to surges in pressure and pulsations from the pump. Their reinforced construction permits hoses to expand slightly and absorb changes in pressure.

Where two diameters of hose are used on the pressure side, the larger diameter, or pressure, hose is at the pump end. It acts as a reservoir and as an accumulator absorbing pulsations. The smaller diameter, or return, hose reduces the effects of kickback from the gear itself. By restricting fluid flow, it also maintains constant back pressure on the pump, which reduces pump noise. If the hose is of one diameter, the gearbox is performing the damping functions internally.

Because of working fluid temperature and adjacent engine temperatures, these hoses must be able to withstand temperatures up to 300°F (149°C). Due to various weather conditions, they must also tolerate subzero temperatures as well. Hose material is specially formulated to resist breakdown or deterioration due to oil or temperature conditions.

Caution: Hoses must be carefully routed away from hot engine manifolds. Power-steering fluid is very flammable. If it comes in contact with hot engine parts, it could start a fire.

Steel tubing and flexible synthetic rubber hosing serve as the arteries and veins of the hydraulic brake system. These brake lines transmit brake fluid pressure from the master cylinder to the wheel cylinders and calipers of the drum and disc brakes.

Fluid transfer from the driver-actuated master cylinder is usually routed through one or more valves and then into the steel tubing and hoses. The design of the brake lines offers quick fluid transfer response with very little friction loss. Engineering and installing the brake lines so they do not wrap around sharp curves is very important in maintaining this good fluid transfer.

Task C.14.9

Determine hose and tubing size.

The size and composition of hose depends largely on its application. Brake line tubing, for example, usually consists of copper-fused double-wall steel tubing in diameters ranging from ⅛ to ⅜ inch (3.175 to 9.525 mm). Some original equipment manufacturer (OEM) brake tubing is manufactured with soft steel strips sheathed with copper. The strips are rolled into a double-wall assembly and then bonded in a furnace at extremely high temperatures. Corrosion protection is often added by tin-plating the tubing.

Assorted fittings are used to connect steel tubing to junction blocks or other tubing sections. The most common fitting is the double or inverted flare. Double flaring is important to maintain the strength and safety of the system. Single flare or sleeve compression fittings may not hold up in the rigorous operating environment of a standard vehicle brake system.

Fittings are constructed of steel or brass. The 37 degree inverted flare or standard flare fitting is the most commonly used coupling. Late model vehicles may use ISO or metric bubble flare fittings.

Never change the style of fitting being used on the vehicle. Replace ISO fittings only with ISO fittings. Replace standard fittings with standard fittings. The metal composition of the fittings must also match exactly. Using an aluminum-alloy fitting with steel tubing may provide a good initial seal, but the dissimilar metals create a corrosion cell that eats away the metal and reduces the connection service life.

Task C.14.10

Recommend proper application and usage of chemicals.

There are many potentially dangerous materials that are encountered in the automotive body shop; materials that can cause bodily harm and property damage if improperly handled. It is important to always follow the manufacturer's suggestions when working with any such material.

Hazardous materials include any material that can cause physical harm or pose a risk to the environment. Hazardous substances are a subset of hazardous materials. These substances pose a threat to waterways and the environment.

The terms *hazardous materials* and *hazardous substances* have specific legal meanings, as identified and regulated by the United States Environmental Protection Agency (EPA). There are laws designed to protect life and the environment from the careless use of hazardous materials. Misuse can result in prosecution and carry severe penalties. It is important to check state and local laws governing the use and disposal of hazardous materials.

There are four basic types of hazardous materials found in the automotive body shop:

    a. Flammable materials—materials that easily catch on fire or explode.

    b. Corrosive materials—materials that will dissolve metals and can damage the skin.

    c. Reactive materials—materials that become unstable and are likely to burn, explode, and/or give off toxic fumes.

d. Toxic materials—materials that cause illness or death from contact, ingestion, and/or inhalation.

Special compounds are available for absorbing oil and cleaning oil spots. Personnel and customers can slip and fall on dirty, oily, or greasy floors. Due to the many hazards encountered in an automotive body shop, signs should be posted to inform visitors and customers that the work areas are for personnel only.

Electrical outlets should be provided throughout the shop, so that there is no need to have extension cords running across the floor.

There is always the possibility of cuts, scrapes, bruises, and pulled muscles in the auto body shop because the technician often works with sharp, heavy sheet metal tools and parts. Gloves should be worn to protect the hands. When lifting heavy items, keep the back straight and lift with the legs.

Everyone in the shop should know how to administer basic first aid. All accidents and injuries must be reported in accordance with posted shop rules and regulations. Emergency numbers should be posted near each telephone in the shop.

D. Vehicle Identification (3 Questions)

Task D.1

Locate and utilize vehicle ID number (VIN).

The standard location for the vehicle identification number (VIN) is attached to the driver side of the instrument panel and is visible through the windshield.

Task D.2

Locate production date.

The production date is the date which the vehicle was assembled. The standard location for this information is on a tag affixed to the driver's door by the door latch, or on the driver's door sill plate.

Task D.3

Locate and utilize component identification data.

Component identification (ID) data are stamped in the casting, or on a tag that is attached to the component. All major components on the vehicle will have ID data attached.

Task D.4

Identify body styles.

The body styles are as follows:
   a. Coupe—a two door vehicle.
   b. Sedan—a four door vehicle.
   c. Station wagon—a vehicle having a passenger compartment which extends to the back of the vehicle.
   d. Regular cab—a cab that has one seating surface and two doors.
   e. Extended cab—a cab, larger than a regular cab, having one and a half seating surfaces. Extended cabs may have two, three, or four doors.
   f. Crew cab—a cab which has two full seating surfaces and four doors.
   g. Step-side bed—a bed that has steps on the outside of the bed and is available in six- and eight-foot lengths.
   h. Style-side bed—a bed having no steps available in six- and eight-foot lengths.

Task D.5

Utilize additional references.

Vehicle build sheets can be a good source of additional information. Vehicle build sheets include the vehicle's identification number (VIN), color, engine size, transmission type, and axle ratio.

Task D.6 **Locate paint code(s).**

The paint code is not part of the vehicle identification number (VIN). It is located on a separate trim tag, called the service parts identification label, which is mounted under the hood or on the driver's door of some vehicles.

E. Cataloging Skills (7 Questions)

Task E.1 **Locate proper catalog.**

Each catalog has, in the upper right-hand corner, the form number, the date of issue, and the number and date of the catalog it replaces or supplements. The cover also identifies the product line and manufacturer and often provides an area where the jobber can stamp his or her name for the convenience of dealer customers.

Task E.2 **Obtain and interpret additional information.**

Every parts specialist will have occasion to refer to catalogs at some point, either to reference additional information, determine exceptions, or find alternative parts.

Sometimes footnotes are difficult to understand when a number of them are squeezed on a page. The footnotes themselves are important. Footnotes should be referenced systematically in order to fully define those applications that have multiple possibilities.

Task E.3 **Utilize additional reference material (technical bulletins, interchange lists, supplements, etc.).**

Bulletins are lists of updated information that manufacturers send to jobbers between issues of their catalogs. They are the primary tool for conducting catalog maintenance, which involves continually updating and revising the catalog racks and making sure counter personnel are using the most up-to-date information from those catalogs.
Bulletins are usually one of the following types:
   a. New item availability bulletins which list items now in stock.
   b. Supersession bulletins, which note part numbers that now supersede previously noted numbers.
   c. Product information bulletins, which contain specific information about a product, such as a manufacturing defect or a unique method of installation.
   d. Technical bulletins, which alert counter personnel to any unusual installation or fit problems or unique maintenance tips.
   e. Correction bulletins, which refer to catalog errors due to mistyping or inaccurately assigned parts numbers.

Task E.4 **Identify catalog terminology and abbreviations.**

Manufacturers usually include additional aids for using their publications such as an abbreviation list which can make for efficient use of the catalog. For example, abbreviations can often have more than one meaning. FWD can mean front-wheel drive or four-wheel drive. OD can mean overdrive or outside diameter. The definitions of these, and other abbreviations, can only be determined by checking the abbreviation list provided.

Task E.5 **Locate index and table of contents.**

The table of contents and index will show the catalog sequence. It will be quite evident if trucks are listed with passenger cars, or if imports are integrated or listed separately. Product additions, possibly resulting from the consolidation of one or more formerly separate catalogs, can be quickly spotted. Instant recognition of several other catalog entities can also be found. These include numerical listings, buyer's guides, progressive size listings, illustrations, installation instructions, competitive interchanges, and more.

Task E.6   **Perform catalog maintenance.**

Bulletins are the primary tool for conducting catalog maintenance, which involves continually updating and revising the catalog racks and making sure counter personnel are using the most up-to-date information from those catalogs.

F. Inventory Management (5 Questions)

Task F.1   **Report lost sales.**

The main benefit of using lost sales reports is the ability to spot trends in the after-market parts business. These reports will show new part numbers that the store does not stock as well as older part numbers that are increasingly in demand. Some part numbers are assured of almost instant demand, such as those found in some of the newer downside models. Others parts will grow in demand, but somewhat more slowly. The store owner or buyer must analyze not only the part numbers that are selling, but also those that the store does not stock.

Task F.2   **Verify incoming and outgoing merchandise.**

As with any delivery, an order for a shipment must be carefully picked, packaged, and documented. A packing list or slip must be included with each shipment. This list contains a detailed description of the items included in the shipment. A separate packing list can be placed in each part of a multiple-part shipment, but often a single comprehensive list is packaged with or secured to the first part of a multiple-part shipment.

Task F.3   **Perform physical inventory.**

Physical inventory should be done once a year. Physical inventory is when all stock is pulled off of the shelves and checked to determine what is actually in the store or shop versus what is in the computer. This is one way to keep the computer system up-to-date and can help you to give the customer faster and more accurate service.

Task F.4   **Report inventory discrepancies.**

After the physical inventory has been completed, the discrepancies must be recorded and put into the computer system. The discrepancies are reported by filling out an inventory discrepancy form. After discrepancies have been entered into the computer system, they should be kept and filed for future reference.

Task F.5   **Perform stock rotation.**

Stock rotation is moving the older stock in front of the newer stock. This prevents the older stock from sitting at the back of the shelf and eventually getting an aged appearance. The stock should be rotated every time that the shelves are stocked. Rotating the product so that the labels are toward the front of the shelf is known as facing. Rearranging the shelves for a neater appearance is done whenever necessary to increase appeal to the customer.

Task F.6   **Handle special orders.**

A special order is placed whenever a customer purchases an item not kept in stock. A part that repeatedly appears on lost sales or special order reports should be considered for stocking status.

Task F.7   **Perform proper core handling (i.e., accepting or declining cores, storage, and return).**

The parts counterperson should be careful when handling cores and accepting core returns from customers. A core charge is a charge that is added when the customer buys a remanufactured or reconditioned part. Core chargers are refunded to the customer when the old defective but rebuildable part is returned. To ensure that customers return their cores, a parts specialist should always require a core charge. There are some parts specialists, however, who believe that charging for the core damages customer relations. While some customers might reliably return a core of good value, there are those who do not return any cores or who return cores of little or no value. There are still others who may return those with a core charge and resell others to core scavengers or scrap metal dealers.

Task F.8

Handle warranty returns.

Additional forms must usually be completed for warranty-return parts. This allows the manufacturers to determine what is wrong with the part and to credit the store for the return. A warranty-return part should never be placed back in the inventory to be sold to another customer.

Task F.9

Determine proper selling unit (each, pair, case, etc.) increment.

The term *each* refers to a single item. *Carton* or *case* refers to a number of parts packaged together and sold as a unit from the supplier. *Pair* refers to two of something. A *set* refers to two or more compatible items. A *kit* usually refers to a group of components used to rebuild a part.

Task F.10

Handle return of broken kits, special-order parts, and exchange parts.

To maintain an adequate supply of good used parts for remanufacturing, a core charge may be added when buying a remanufactured part. This charge is refunded when the customer returns the used part. Freight charges may be added to emergency order parts to cover the cost of special transportation to the store. It is not uncommon to add a long-distance phone charge for special phone orders. Some parts may have a restocking fee that is charged to the customer when the parts are returned for a refund.

G. Merchandising (3 Questions)

Task 1 G.1

Understand display strategy.

The best place for a display is in an open area near the front of the store or department. This ensures that the display will catch the customer's eye as soon as he or she enters. The display should be large and noticeable and it must look appealing to the customer. This is usually accomplished by building the display out of the product being sold. Often the manufacturer of a product will send a display kit that can be assembled in the store and which may include banners, signs, and shelves. Displays are also often built at the end of an isle where the product is located so the customer will notice the product without entering the isle. Displays can be set up with the intent of the customer taking the actual item from the display. The display should be set up with care to keep it from falling apart as items are removed. A display also needs to be maintained for a neat appearance and to keep it looking appealing to the customer.

Task 2 G.2

Display pricing.

A normal practice of display pricing is to display an item that may be specially priced for that week in an eye-catching display in an open part of the store or department, along with a noticeable sign that displays the special pricing. The product that is specially priced may still remain in its assigned shelf position but this additional display should let customers know this item is on special. A customer is more likely to notice a display with a posted price rather than notice an item on the shelf in its normal position. This kind of display and pricing is also used for the introduction of new items. A vendor may also choose to put up a display if overstocked in a certain item but choose not to lower the price. This will help promote the item and possibly lead to more sales.

Display pricing can also be used to bring customers to the department or store by the distribution of fliers with an explanation and the price, by advertising on a billboard with the price, or signs in a window with the price.

Task G.3

Inspect and maintain shelf quantities and conditions.

In order to keep shelves and items stocked and looking appealing to a customer the shelves must be well maintained. The shelves, as well as the items on the shelves, should not be dirty or dusty. Dusty items will not sell well because nobody wants to buy anything that they may feel is old and has been sitting around forever. The items on the shelves should be replenished on a regular basis to encourage repeat customers to count on you to have the items they need at all times. If shelves are messy and items are not easily found, or shelves are always empty from not restocking, often enough this will discourage a sale or a customer from revisiting your store or department again. The next item on a shelf should be pulled forward when the one in front of it is sold. This will give your shelf the appearance of being full and items will be easier to find for the next customer. When items are restocked the new items should be put behind the old ones. This will help keep dust from building up on items.

A shelf should be labeled to some extent to let people know which items go where. Displaying a price on a shelf where the item goes will help keep shelves in order because a customer will not have to take the item off the shelf to see how the price may differ from a similar item next to it. A customer may not always return the item to the space where it was removed, causing an unorganized appearance. Returned items should not be put back on the shelf if they are opened as this will give them a used appearance and make the customer uneasy about purchasing them. Items that are returned unopened can be returned to the shelf if they appear to be in good order.

Task G.4

Identify impulse, seasonal, and related items.

A seasonal item is something that a parts department may not normally stock during certain times of the year. Some seasonal items are stocked throughout the entire year but not in large quantities. An example of one of these items is windshield washer fluid, which is stocked more heavily in the winter season than the summer season but is still sold throughout the year. Another such example is car batteries, which are stocked all year even though battery failures occur more often during the colder parts of the year. Some other items used year-round but stocked more heavily in the winter include antifreeze, winter windshield wipers, and gasoline additives to prevent gas line freeze. Some items more heavily stocked in the summer include air conditioning refrigerant, windshield sun shades, and wax or polish. An impulse item is a product that the customer buys on the spur of the moment to fill a "want" rather than a "need" for the item. Some examples of impulse items include vehicle appearance-enhancing items such as pinstriping kits, chrome accessories, high-flow air filters to improve vehicle performance, and interior items like extra cup holders or seat covers.

Task G.5

Utilize sales aides.

Most merchandisers agree that the best location for impulse items is at a point between the customer's chest and eye level. By careful planning, the placement of products at appealing locations throughout the store greatly increases sales. The average customer is very sensitive to the prices of commonly purchased merchandise. While some products, such as motor oil, may be a high volume item for the merchant, they are seldom high profit items. Changing motor oil is the most popular task performed by the do-it-yourselfer (DIYer). Also, it is the reason most vehicles visit a garage. Motor oil is a product that is often sold at near cost as a price leader to attract customers.

5 Sample Test for Practice

Sample Test

Please note the letter and number in parentheses following each question. They match the overview in section 4 that discusses the relevant subject matter. You may want to refer to the overview using this cross-referencing key to help with questions posing problems for you.

1. Parts Specialist A says a 10 percent discount for $25.00 is $2.50. Parts Specialist B says a part bought for $10.00 and sold for $15.00 will generate a 50 percent profit. Who is right?
   A. A only
   B. B only
   C. Both A and B
   D. Neither A nor B   (A.1)

2. If a customer returns a $100.00 part and is charged a 5 percent restocking fee, how much money is returned to the customer?
   A. $5.00
   B. $95.00
   C. $100.00
   D. $105.00   (A.2)

3. Parts Specialist A says that 350 cubic inches is about 5.0 liters. Parts Specialist B says that 390 cubic inches is about 6.6 liters. Who is right?
   A. A only
   B. B only
   C. Both A and B
   D. Neither A nor B   (A.3)

4. Which of these numbers would appear first in an alphanumeric listing?
   A. 369482A
   B. 378654C
   C. 369482B
   D. 388426A   (A.4)

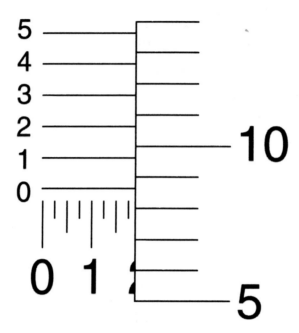

5. The 0–1 micrometer reading in the illustration is:
   A. 0.108 inch.
   B. 0.184 inch.
   C. 0.255 inch.
   D. 0.288 inch.   (A.5)

6. Parts Specialist A says that charge accounts are for large sales. Parts Specialist B says that you must be familiar with the charge policies before you make the transaction. Who is right?
   A. A only
   B. B only
   C. Both A and B
   D. Neither A nor B   (A.7)

7. Parts Specialist A says the ability to communicate is essential. Parts Specialist B says that interacting with all people is easy. Who is right?
   A. A only
   B. B only
   C. Both A and B
   D. Neither A nor B   (A.8)

8. Parts Specialist A says a counterperson has the responsibility for keeping the entire store looking professional. Parts Specialist B says the store should have someone hired to clean up the shop. Who is right?
   A. A only
   B. B only
   C. Both A and B
   D. Neither A nor B   (A.9)

9. Parts Specialist A says it is better to ask for help than to sell the customer the wrong parts. Parts Specialist B says that an experienced employee may be asked to help train a new employee. Who is right?
   A. A only
   B. B only
   C. Both A and B
   D. Neither A nor B (A.10)

PORT 2	PARTS-J3	INVOICING	10OCT00 1337

Invoice : 12223
Customer: 99999
Name: CASH RETAIL
Address:
City,St Zip:
Home Phone:

Sale Type: CASH

Parts: 158.40
Freight: 0.00
Tax: 9.50
Total Invoice: 167.90
Backorder Amount: 0.00

Emp: 117   Sales Person:-   Ship Via: carry   PO: none   B/L:-

Part No.	Description	Bin	O.H.	Cost	Sale	Ext. sale	Q.S.	# A	O.O.	PM
23196550	WIPER MOTOR	245	11	62.36	87.30	87.30	1			
74073158	WIPER INSERTS	108	3	5.82	8.15	16.30	2			
21646146	WIPER SWITCH	218	8	37.38	52.33	52.33	1			
7148887	GROMMETS	13B	5	0.31	0.43	1.74	4			M
27289127	25 A FUSE	2A1	1	0.52	0.73	0.73	1			

F1=Help   F3=Save   F4=Cancel   F8=Print   F10=login

10. According to the invoice, what would be the total amount returned to the customer if the wiper motor was returned? Your shop has a policy to accept all returns providing they were not installed, are in their original package, and are not damaged.
    A. $87.30
    B. $62.36
    C. $92.54
    D. $66.10 (A.6)

11. A parts specialist should always:
    A. get help before heavy lifting.
    B. tell someone else to move heavy objects.
    C. size up the task before lifting an object.
    D. use a hand truck when lifting an object. (A.11)

12. The Environmental Protection Agency (EPA) and the Occupational Safety and Health Administration (OSHA) give strict guidelines on all of the following chemicals EXCEPT:
    A. solvents.
    B. floor soaps.
    C. cutting oils.
    D. caustic cleaning compounds. (A.12)

13. Parts Specialist A says that large items should be displayed on the top shelf so the customer can see them. Parts Specialist B says expensive items should be displayed behind the counter in locked cases. Who is right?
    A. A only
    B. B only
    C. Both A and B
    D. Neither A nor B   (A.13)

14. A do-it-yourselfer (DIYer) usually needs:
    A. a professional technician's assistance.
    B. used or rebuilt parts and accessories.
    C. rebuilt engines and transmissions.
    D. advice on parts and methods.   (B.1)

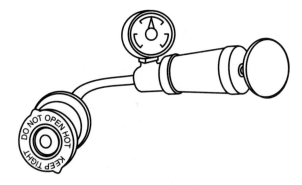

15. Parts Specialist A says the setup is used to check a radiator cap's ability to remain sealed as temperature changes. Parts Specialist B says the tester is checking the cap's ability to remain sealed as pressure increases. Who is right?
    A. A only
    B. B only
    C. Both A and B
    D. Neither A nor B   (C.2.4)

16. Parts Specialist A says that "What do you need?" is the best opening question to ask a customer. Parts Specialist B says that "May I help you?" is the best opening question to ask a customer. Who is right?
    A. A only
    B. B only
    C. Both A and B
    D. Neither A nor B   (B.2)

17. Parts Specialist A says that you should always establish who is to blame for a problem. Parts Specialist B says that you should raise your voice to stay above the angry customer's voice. Who is right?
    A. A only
    B. B only
    C. Both A and B
    D. Neither A nor B   (B.4)

18. Parts Specialist A says that the parts specialist should wait until it is the customer's turn before establishing eye contact. Parts Specialist B says that you should never talk to the customer while you are looking in the parts books. Who is right?
    A. A only
    B. B only
    C. Both A and B
    D. Neither A nor B (B.5)

19. The phone rings while the parts specialist is helping a customer at the counter and there is no one else available to answer the phone. What should the parts specialist do? Parts Specialist A says to politely ask the counter customer to wait while you answer the phone. Parts Specialist B says to yell for assistance from another parts specialist. Who is right?
    A. A only
    B. B only
    C. Both A and B
    D. Neither A nor B (B.6)

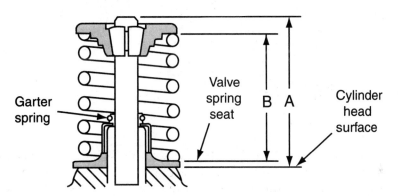

20. Which of the following best describes what dimension A in the drawing represents?
    A. Valve spring free height
    B. Valve spring installed height
    C. Valve stem height
    D. Valve retainer to seat clearance (C.1.4)

21. Parts Specialist A says that the appearance of the store as well as the appearance of the counter personnel influence the customer. Parts Specialist B says that a clean, neat parts specialist can influence the customer. Who is right?
    A. A only
    B. B only
    C. Both A and B
    D. Neither A nor B (B.8)

22. Parts Specialist A says that selling related parts can boost profits by as much as 30 percent. Parts Specialist B says that you should always promote stock that is not selling well. Who is right?
    A. A only
    B. B only
    C. Both A and B
    D. Neither A nor B (B.9)

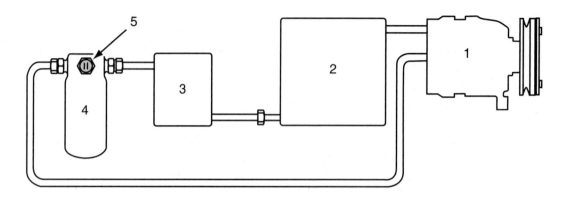

23. While discussing the air conditioning system drawn in the figure, Parts Specialist A says the component labeled 3 is the evaporator. Parts Specialist B says the component labeled 1 is the compressor. Who is right?
    A. A only
    B. B only
    C. Both A and B
    D. Neither A nor B   (C.12.1)

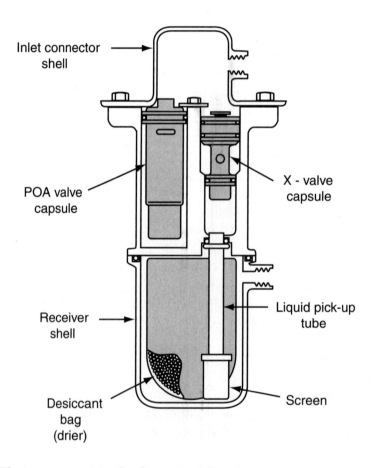

24. The component in the figure is a(n):
    A. accumulator.
    B. receiver/drier.
    C. VIR (valves in receiver) unit.
    D. refrigerant filter and valve assembly.   (C.12.1)

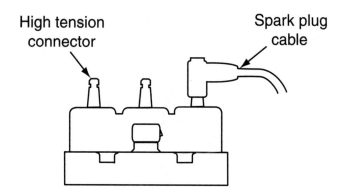

High tension connector

Spark plug cable

25. Parts Specialist A says the part shown in the figure is the ignition coil pack for at least three cylinders. Parts Specialist B says the part shown is an ignition coil pack for a distributorless ignition system. Who is right?
    A. A only
    B. B only
    C. Both A and B
    D. Neither A nor B   (C.4.1)

26. All of the following components are in the fuel system EXCEPT:
    A. the fuel pump.
    B. the fuel tank.
    C. the carburetor.
    D. fuel tank hangers.   (C.3.1)

27. Parts Specialist A says that an EGR valve has a poppet valve and a diaphragm. Parts Specialist B says that EGR valves are used for the cruise control system. Who is right?
    A. A only
    B. B only
    C. Both A and B
    D. Neither A nor B   (C.6.2)

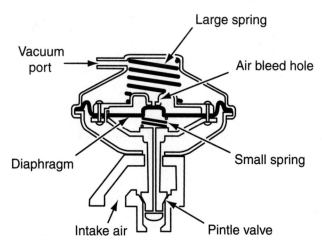

Large spring

Vacuum port

Air bleed hole

Diaphragm

Small spring

Intake air

Pintle valve

28. While discussing the figure, Parts Specialist A says the part is a major component of the emission control system. Parts Specialist B says the part is an EGR valve. Who is right?
    A. A only
    B. B only
    C. Both A and B
    D. Neither A nor B   (C.6.1)

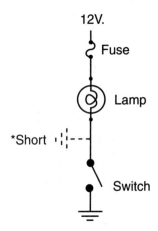

29. Which of the following statements is true about the short shown in the drawing?
    A. The fuse would blow.
    B. The lightbulb would burn out quickly.
    C. The lamp would remain on at all times.
    D. The switch would heat up due to arcing across the contacts.   (C.13.4)

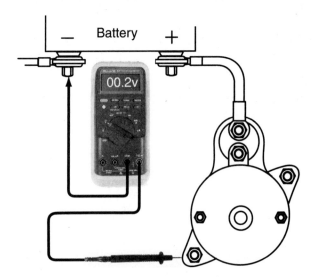

30. Parts Specialist A says the meter hookup shown in the drawing is measuring the available voltage to the starter motor. Parts Specialist B says the reading of 0.2 volts shows the starter is bad. Who is right?
    A. A only
    B. B only
    C. Both A and B
    D. Neither A nor B   (C.13.4)

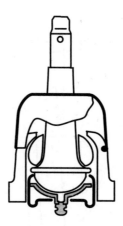

31. The suspension part shown in the figure is a(n):
    A. tie rod end.
    B. ball joint.
    C. idler arm.
    D. wheel spindle. (C.11.1)

32. Parts Specialist A says that the suspension keeps the vehicle in contact with the ground. Parts Specialist B says that the suspension helps to cushion the vehicle from road shock. Who is right?
    A. A only
    B. B only
    C. Both A and B
    D. Neither A nor B (C.11.2)

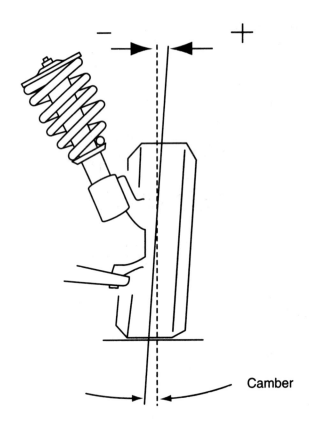

Camber

33. Referring to the drawing, Parts Specialist A says a bent strut will cause this alignment angle to be out of specs. Parts Specialist B says if this angle is wrong, excessive tire wear will result. Who is right?

    A. A only

    B. B only

    C. Both A and B

    D. Neither A nor B   (C.11.4)

34. Parts Specialist A says that the starter motor turns a flywheel mounted on the rear of the crankshaft. Parts Specialist B says that when the ignition switch is turned to START, there is electrical power directly to the alternator. Who is right?

    A. A only

    B. B only

    C. Both A and B

    D. Neither A nor B   (C.13.2)

35. Bolt dimensions are determined by the:

    A. distance across the flats of the hex head.

    B. size wrench that fits on the bolt head.

    C. distance across the points of the hex head.

    D. diameter and pitch of the bolt threads.   (C.14.1)

36. Parts Specialist A says that the location of the vehicle identification number (VIN) tag is on the driver's side inner fender. Parts Specialist B says that the vehicle identification number (VIN) tag is visible through the driver's door window. Who is right?

    A. A only

    B. B only

    C. Both A and B

    D. Neither A nor B   (D.1)

37. The production date tag will be found:
    A. inside the trunk lid.
    B. on the passenger side rear door or jamb.
    C. on the driver's door or sill plate.
    D. under the hood on the passenger side inner fender. (D.2)

38. All of the following have a tag similar to the one shown in the figure EXCEPT the:
    A. engine.
    B. transmission.
    C. driveshaft.
    D. axle. (D.3)

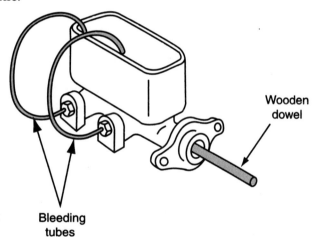

Wooden dowel

Bleeding tubes

39. While discussing the setup shown in the drawing, Parts Specialist A says the procedure is called bench bleeding. Parts Specialist B says the procedure is bench bleeding and is done to eliminate the need to bleed the entire brake system after a master cylinder is installed in a vehicle. Who is right?
    A. A only
    B. B only
    C. Both A and B
    D. Neither A nor B (C.10.4)

40. Parts Specialist A says that the paint code is in the vehicle identification number (VIN). Parts Specialist B says that the paint code can be found by looking in the owner's manual. Who is right?
    A. A only
    B. B only
    C. Both A and B
    D. Neither A nor B (D.6)

41. Parts Specialist A says that each catalog has a form number. Parts Specialist B says that the cover to each catalog should identify the product and manufacturer. Who is right?

    A. A only

    B. B only

    C. Both A and B

    D. Neither A nor B   (E.1)

42. Parts Specialist A says that footnotes can be very important for getting the correct part. Parts Specialist B says that footnotes are provided only for special order parts. Who is right?

    A. A only

    B. B only

    C. Both A and B

    D. Neither A nor B   (E.2)

43. Three types of fittings used on automobile hoses and tubes are:

    A. flare, compression, and tubing.

    B. pipe, flare, and steel.

    C. tubing, steel, and pipe.

    D. flare, compression, and pipe.   (C.14.5)

44. All of the following are types of bulletins EXCEPT:

    A. new item availability bulletins.

    B. delivery bulletins.

    C. product information bulletins.

    D. correction bulletins.   (E.3)

ABS	- Anti-lock Brake System	FWD	- Front Wheel Drive	P	- Passenger's Side
AC	- Air Conditioning	I.D.	- Inside Diameter	Qty.	- Quantity
AT	- Automatic Transmission	Man	- Manual	RWD	- Rear Wheel Drive
ATC	- Automatic Temperature Control	min.	- Minimum	Veh.	- Vehicle
		mm	- Millimeters	w/	- With
Auto	- Automatic	Max.	- Maximum	w/o	- Without
cid	- Cubic Inch Displacement	MT	- Manual Transmission	3AUT	- 3 Speed Automatic Transmission
cc	- Cubic Centimeters	NA	- Not Available	4AUT	- 4 Speed Automatic Transmission
Comp.	- Competitor	No.	- Number	4MAN	- 4 Speed Manual Transmission
C.V.	- Constant Velocity	NR	- Not Required	4WD	- Four Wheel Drive
D	- Driver's Side	O.D.	- Outside Diameter	5MAN	- 5 Speed Manual Transmission
dia.	- Diameter	O.E.	- Original Equipment	6MAN	- 6 Speed Manual Transmission
exc	- Except				

45. In the figure, the abbreviation AC means:

    A. automatic controls.

    B. asbestos contamination.

    C. air conditioning.

    D. air compressor.   (E.4)

46. A customer enters an automotive parts store only to browse. Which of these items is this customer most likely to buy?

    A. A pair of windshield wipers

    B. Coolant flush for a cooling system

    C. A transmission service kit

    D. A vehicle pinstripe kit   (G.4)

47. Parts Specialist A says that there is no reason to rotate stock on items that are not perishable. Parts Specialist B says that if the price is displayed on a shelf it may help the appearance of the store or department. Who is right?
    A. A only
    B. B only
    C. Both A and B
    D. Neither A nor B   (G.3)

48. All of the following are true about display pricing EXCEPT:
    A. Display pricing is often used for the introduction of new items.
    B. Display pricing is often used for special sale items.
    C. The customer is more likely to notice the price and product in a display rather than on a shelf.
    D. If an item is in a display then it will always be on sale.   (G.2)

49. A display arrangement must be:
    A. no higher than four feet.
    B. in an open area to be effective.
    C. near where the product usually is.
    D. set up by a display person.   (G.1)

50. A customer asks for four liters of oil. How many quarts should the parts specialist give the customer?
    A. Two
    B. Four
    C. Five
    D. Six   (A.3)

51. Parts Specialist A says that an invoice should be completed for each sale. Parts Specialist B says that back orders are the type of orders made to maintain the present stock in the store. Who is right?
    A. A only
    B. B only
    C. Both A and B
    D. Neither A nor B   (A.14)

52. Parts Specialist A says that talking down to customers could affect the store's sales. Parts Specialist B says that the attitude of the parts specialist could have a great effect on the customer's confidence. Who is right?
    A. A only
    B. B only
    C. Both A and B
    D. Neither A nor B   (B.3)

53. Parts Specialist A says that some stores have policies that will not allow the customer to return parts that have been installed. Parts Specialist B says that most do-it-yourselfers have good enough diagnostic skills to replace the correct part. Who is right?
    A. A Only
    B. B Only
    C. Both A and B
    D. Neither A nor B   (B.7)

54. A customer asks for a set of brake shoes and the parts specialist has four different kinds available. What should the parts specialist do next?

    A. Sell the customer the best set of brake shoes.

    B. Sell the customer the most expensive set of brake shoes.

    C. Tell the customer about the different kinds of brake shoes and let the customer make the decision.

    D. Sell the customer the cheapest set of brake shoes because the vehicle is old and not worth very much.   (B.10)

55. Parts Specialist A says that a parts specialist should never push a sale. Parts Specialist B says that it is part of a parts specialist's job to identify whether the customer is there to look around or there to buy something. Who is right?

    A. A only

    B. B only

    C. Both A and B

    D. Neither A nor B   (B.11)

56. Parts Specialist A says that if you are helping a customer at the counter and the telephone rings, you should just push the hold button. Parts Specialist B says that you should never interrupt activities with a customer at the counter by answering the telephone. Who is right?

    A. A only

    B. B only

    C. Both A and B

    D. Neither A nor B   (B.12)

57. A customer asks for brake shoes. While looking up the right application for the brake shoes the parts specialist should:

    A. not take the time to answer the phone.

    B. finish lunch (if during lunchtime).

    C. talk to other counter personnel.

    D. suggest related items for the brake repair.   (B.13)

58. Parts Specialist A says that if a DIY customer asks for brake shoes it is not necessary to ask if brake fluid is needed. Parts Specialist B says that suggesting related items is only done when talking to a trained repair technician. Who is right?

    A. A only

    B. B only

    C. Both A and B

    D. Neither A nor B   (B.14)

59. A customer comes into the store carrying a set of jumper cables. The customer asks for a starter motor for his late model vehicle. What should the parts specialist do next?

    A. Look up the starter motor for the vehicle.

    B. Ask the customer what symptoms suggested the need for a starter.

    C. Refer the customer to a good automotive repair facility.

    D. Ask another parts specialist what starters fit the customer's vehicle.   (B.15)

60. Closing the sale means:

    A. getting a commitment from the customer.

    B. taking the customer's money.

    C. looking up the correct part.

    D. handing the customer their change and receipt.   (B.16)

61. Parts Specialist A says that the vehicle build sheet can indicate the vehicle color. Parts Specialist B says that the vehicle build sheet can indicate the vehicle engine size. Who is right?
    A. A only
    B. B only
    C. Both A and B
    D. Neither A nor B   (D.5)

62. Parts Specialist A says that the table of contents will show the catalog sequence. Parts Specialist B says that the index will show the catalog sequence. Who is right?
    A. A only
    B. B only
    C. Both A and B
    D. Neither A nor B   (E.5)

63. Parts Specialist A says that bulletins are the primary tool for conducting catalog maintenance. Parts Specialist B says that bulletins are not very useful and that posting them is a waste of effort. Who is right?
    A. A only
    B. B only
    C. Both A and B
    D. Neither A nor B   (E.6)

64. Parts Specialist A says that a packing list should be included with all deliveries. Parts Specialist B says that the packing list gives directions for where to deliver the package. Who is right?
    A. A only
    B. B only
    C. Both A and B
    D. Neither A nor B   (F.2)

65. Physical inventory should be done:
    A. every day.
    B. weekly.
    C. monthly.
    D. annually.   (F.3)

66. Physical inventory discrepancies are reported by:
    A. telling the store manager.
    B. calling the jobber and letting them know about the discrepancy.
    C. filling out an inventory discrepancy form.
    D. letting the other parts specialists know about the discrepancies.   (F.4)

67. Parts Specialist A says that stock rotations are not necessary. Parts Specialist B says that stock rotation may be done annually. Who is right?
    A. A only
    B. B only
    C. Both A and B
    D. Neither A nor B   (F.5)

68. A special order is:
    A. the order that is made every week.
    B. a part ordered because the store does not keep it in stock.
    C. an order made because a vehicle is inoperative and needs to be returned to service as soon as possible.
    D. an order placed for the parts manager or store owner.   (F.6)

69. A core charge is:
    A. a different name for a state sales tax.
    B. the charge added by the store above the usual markup.
    C. a charge added when a customer buys a remanufactured part.
    D. the charge for restocking a returned part.   (F.7)

70. Parts Specialist A says that an additional form must be filled out for warranty returns. Parts Specialist B says that a part returned for warranty should not be put back on the shelf and sold to another customer. Who is right?
    A. A only
    B. B only
    C. Both A and B
    D. Neither A nor B   (F.8)

71. A case refers to:
    A. a package of five parts.
    B. the number of parts packaged together and sold as a unit.
    C. the number of parts in a display box.
    D. a package of twenty pieces.   (F.9)

72. A restocking fee may be charged to a customer for:
    A. buying too many of one type of a part.
    B. having the brake rotors turned.
    C. not buying enough parts each month.
    D. returning a part for a refund.   (F.10)

73. Parts Specialist A says that there are no benefits to using lost sales reports. Parts Specialist B says that most stores analyze not only the parts that they are selling, but parts that the store does not sell so well. Who is right?
    A. A only
    B. B only
    C. Both A and B
    D. Neither A nor B   (F.1)

74. Parts Specialist A says excessive flywheel runout may cause grabbing or erratic clutch operation. Parts Specialist B says the pressure plate should always be reinstalled in the original position on the flywheel. Who is right?
    A. A only
    B. B only
    C. Both A and B
    D. Neither A nor B   (C.1.3)

75. A port fuel-injected engine has a steady "puff" noise in the exhaust with the engine idling. Which of the following may cause this problem?
    A. A burned exhaust valve
    B. Excessive fuel pressure
    C. A restricted fuel return line
    D. A sticking fuel pump check valve   (C.1.4)

76. The following are normal oil pump component measurements EXCEPT:
    A. inner rotor diameter.
    B. clearance between the rotors.
    C. inner and outer rotor thickness.
    D. outer rotor to housing clearance. (C.2.3)

77. Water pump bearings are being discussed. Parts Specialist A says a defective water pump bearing may cause a growling noise when the engine is idling. Parts Specialist B says the water pump bearing may be ruined by coolant leaking past the pump seal. Who is right?
    A. A only
    B. B only
    C. Both A and B
    D. Neither A nor B (C.2.4)

78. Fuel system problems are being discussed. Parts Specialist A says an intake manifold vacuum leak may cause a cylinder misfire with the engine idling. Parts Specialist B says an intake manifold vacuum leak may cause a cylinder misfire during hard acceleration. Who is right?
    A. A only
    B. B only
    C. Both A and B
    D. Neither A nor B (C.3.3)

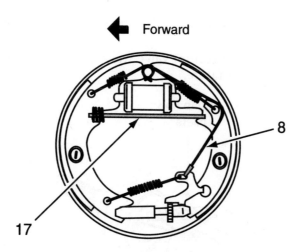

79. Parts Specialist A says the part labeled 8 in the drawing is part of the self-adjusting function of the brake shoe assembly. Parts Specialist B says the part labeled 17 keeps the two brake shoes aligned while they expand and retract. Who is right?
    A. A only
    B. B only
    C. Both A and B
    D. Neither A nor B (C.10.2)

80. The following statements about distributor advances are true EXCEPT:
    A. The vacuum advance controls spark advance in relation to engine load.
    B. The mechanical advance controls spark advance in relation to engine rpm.
    C. The mechanical advance rotates the reluctor in the opposite direction to shaft rotation.
    D. The vacuum advance rotates the pickup plate in the opposite direction to shaft rotation. (C.4.3)

81. A 12V test light connected from the negative primary coil terminal to ground flashes while cranking the engine, but a test spark plug does not fire when connected from the coil secondary wire to ground. Parts Specialist A says the distributor cap and rotor may be defective. Parts Specialist B says the coil may be defective. Who is right?

    A. A only

    B. B only

    C. Both A and B

    D. Neither A nor B   (C.4.4)

82. Parts Specialist A says the positive crankcase ventilation (PCV) system may draw unfiltered air into the engine through a leaking rocker arm cover gasket. Parts Specialist B says the PCV system may draw unfiltered air into the engine through a loose oil filler cap. Who is right?

    A. A only

    B. B only

    C. Both A and B

    D. Neither A nor B   (C.5.3)

83. All of the following statements regarding manifold heat control valves are true EXCEPT:

    A. A manifold heat control valve improves fuel vaporization in the intake manifold especially when the engine is cold.

    B. A manifold heat control valve stuck in the closed position causes a loss of engine power.

    C. A manifold heat control valve stuck in the open position may cause an acceleration stumble.

    D. A manifold heat control valve stuck in the closed position reduces intake manifold temperature.   (C.5.4)

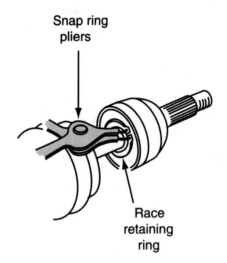

Snap ring
pliers

Race
retaining
ring

84. Which of the following statements about the part shown in the figure is NOT true?

    A. The part is an outboard CV joint.

    B. To remove the part, the axle shaft must be removed first.

    C. The part is typically permanently lubricated.

    D. A boot is installed over the part to prevent moisture and dirt from damaging it.   (C.9.4)

85. When the ignition switch is turned on, 12V is supplied to one terminal on the secondary air injection diverter (AIR Diverter) and secondary air injection bypass (AIR Bypass) solenoids. The air is bypassed normally to the atmosphere for a few seconds when the engine is started. When this bypass mode is completed, the airflow is always directed downstream. The voltage on the powertrain control module (PCM) side of the AIR Bypass solenoid is 0.2V, and the voltage on the PCM side of the AIR Diverter solenoid is 12V. Parts Specialist A says the wire from the AIR Diverter solenoid to the PCM may have an open circuit. Parts Specialist B says the PCM may not be grounding the circuit from the AIR Diverter solenoid. Who is right?
    A. A only
    B. B only
    C. Both A and B
    D. Neither A nor B   (C.6.4)

86. The shift lever adjustment usually is performed with the transmission in:
    A. neutral.
    B. first gear.
    C. second gear.
    D. reverse gear.   (C.7.3)

87. A five-speed manual transaxle has a growling and rattling noise in third gear only. The cause of this noise could be worn:
    A. or chipped teeth on the third speed gear on the input shaft.
    B. or chipped dog teeth on the third gear on the input shaft.
    C. dog teeth on the third speed synchronizer blocking ring.
    D. threads in the cone area of the third speed blocking ring.   (C.7.4)

88. An automatic transmission has a whining noise that occurs in all gears while the vehicle is being driven. This noise is also present with the engine running and the vehicle stopped. Parts Specialist A says the rear planetary gearset may be defective. Parts Specialist B says the oil pump may be defective. Who is right?
    A. A only
    B. B only
    C. Both A and B
    D. Neither A nor B   (C.8.3)

89. Shift linkage adjustments are being discussed. Parts Specialist A says that improper shift linkage adjustments may cause premature transmission clutch failure. Parts Specialist B says that improper shift linkage adjustments may cause higher than normal fluid pressure. Who is right?
    A. A only
    B. B only
    C. Both A and B
    D. Neither A nor B   (C.8.4)

90. A front-wheel-drive car has a clunking noise while decelerating. Parts Specialist A says this noise may be caused by a worn inner drive axle joint. Parts Specialist B says this noise may be caused by a worn front wheel bearing. Who is right?
    A. A only
    B. B only
    C. Both A and B
    D. Neither A nor B   (C.9.3)

91. All of the following statements about differential case and ring gear assembly removal and replacement are true EXCEPT:
    A. The ring gear runout should be measured before removal of the case and ring gear assembly.
    B. The case side play should be measured before removal of the case and ring gear assembly.
    C. The side bearing caps should be marked in relation to the housing before removal of the case and ring gear assembly.
    D. The side bearings should be clean and dry before installation of the case and ring gear assembly.   (C.9.4)

92. Brake lines and hoses are being discussed. Parts Specialist A says a damaged brake line may be repaired with a short piece of line and compression fittings. Parts Specialist B says the necessary brake line bends should be made with a tube-bending tool. Who is right?
    A. A only
    B. B only
    C. Both A and B
    D. Neither A nor B   (C.10.3)

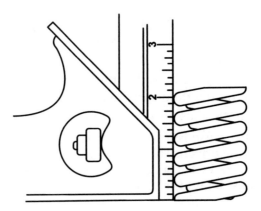

93. While discussing the setup shown in the figure, Parts Specialist A says the free-standing height of the valve spring is being measured. Parts Specialist B says the spring is being checked for squareness. Who is right?
    A. A only
    B. B only
    C. Both A and B
    D. Neither A nor B   (C.1.3)

94. When one side of the front or rear bumper is pushed downward with considerable weight and then released, the bumper makes two free upward bounces before the vertical chassis movement stops. This action indicates:
    A. a defective shock absorber.
    B. a weak coil spring.
    C. a broken spring insulator.
    D. a worn stabilizer bushing.   (C.11.3)

95. A customer complains about harsh riding on the front suspension, excessive steering effort, and rapid steering wheel return. Parts Specialist A says the front suspension may have excessive positive caster. Parts Specialist B says the included angle may be more than specified on the front suspension. Who is right?

   A. A only

   B. B only

   C. Both A and B

   D. Neither A nor B   (C.11.4)

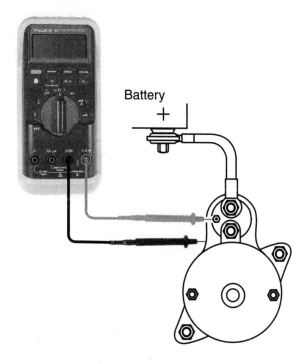

Battery
+

96. Which of the following describes what the meter setup in the figure is measuring?

   A. Voltage drop across the ground circuit

   B. Available voltage at the starter

   C. Voltage drop of the solenoid

   D. Voltage at the battery   (C.13.2)

97. Parts Specialist A says restricted refrigerant passages in the evaporator may cause frosting of the evaporator outlet pipe. Parts Specialist B says restricted refrigerant passages in the evaporator may cause much higher than specified low-side pressures. Who is right?

   A. A only

   B. B only

   C. Both A and B

   D. Neither A nor B   (C.12.4)

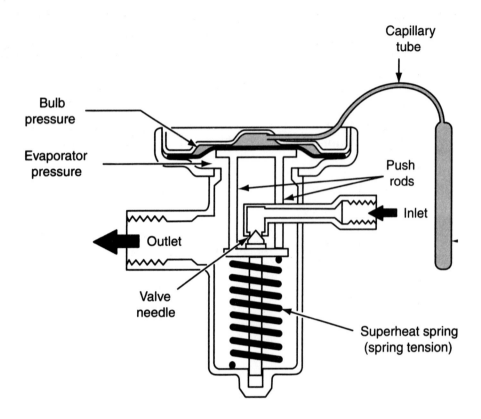

98. While discussing the valve shown in the drawing, Parts Specialist A says the remote bulb must be attached to a refrigerant line in order to accurately control flow to the evaporator. Parts Specialist B says the valve above controls the refrigerant flow in order to maintain proper cooling. Who is right?

    A. A only
    B. B only
    C. Both A and B
    D. Neither A nor B   (C.12.2)

99. When testing a diode, connect the ohmmeter leads across the diode and then reverse the leads. A satisfactory diode provides:

    A. two high meter readings.
    B. one high one low meter reading.
    C. two low meter readings.
    D. a meter reading of 0 Ohms and 10 Ohms.   (C.13.4)

100. Parts Specialist A says that a good location for merchandise to attract impulse sales is on a shelf level that is between the knees and waist of the customer. Parts Specialist B says a good location for merchandise to attract impulse sales is on a shelf that is just above the customer's eye level. Who is right?

    A. A only
    B. B only
    C. Both A and B
    D. Neither A nor B   (G.5)

101. A liquid cooling system is preferred for all of the following reasons EXCEPT:
    A. liquid cooling systems are less noisy than air cooling systems.
    B. liquid cooling systems are less likely to warp metal engine parts than air cooling systems.
    C. liquid cooling systems operate more efficiently than air cooling systems.
    D. liquid cooling systems provide heated coolant for the heater.   (C.2.1)

102. Parts Specialist A says fuel must be atomized for proper combustion. Parts Specialist B says that air is mixed with the fuel for atomization. Who is right?
    A. A only
    B. B only
    C. Both A and B
    D. Neither A nor B   (C.3.2)

103. Parts Specialist A says there is more than one coil in a distributorless ignition system (DIS). Parts Specialist B says the coil synchronizes the ignition module in relation to the crankshaft position. Who is right?
    A. A only
    B. B only
    C. Both A and B
    D. Neither A nor B   (C.4.2)

104. The following are all part of the exhaust system EXCEPT:
    A. the muffler.
    B. the resonator.
    C. the catalytic converter.
    D. the reverberator.   (C.5.2)

105. What agency establishes automotive emissions standards?
    A. EGR
    B. EPA
    C. EEC
    D. EFE   (C.6.1)

106. Parts Specialist A says that pressing the clutch pedal engages the clutch. Parts Specialist B says the clutch pressure plate is operated by a clutch release bearing. Who is right?
    A. A only
    B. B only
    C. Both A and B
    D. Neither A nor B   (C.7.2)

107. Parts Specialist A says that the torque converter provides fluid coupling in the transmission. Parts Specialist B says that the torque converter allows maximum slippage at engine idle speed. Who is right?
    A. A only
    B. B only
    C. Both A and B
    D. Neither A nor B   (C.8.2)

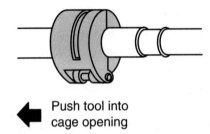

Push tool into
cage opening

108. While discussing the tool shown in the drawing, Parts Specialist A says this tool is required to join two ends of a refrigerant line together. Parts Specialist B says this tool can be used to cut the line so a defective section could be removed. Who is right?
A. A only
B. B only
C. Both A and B
D. Neither A nor B   (C.12.3)

109. Parts Specialist A says the rear brakes provide about 40 percent of the braking capacity if the front brakes are in good order. Parts Specialist B says that the rear brakes will provide 60 percent of the braking capacity if the front brakes have failed. Who is right?
A. A only
B. B only
C. Both A and B
D. Neither A nor B   (C.10.2)

110. Four-wheel steering is being discussed. Parts Specialist A says four-wheel steering is not yet available for practical application. Parts Specialist B says four-wheel steering is only available in sport-utility vehicles (SUVs). Who is right?
A. A only
B. B only
C. Both A and B
D. Neither A nor B   (C.11.1)

111. Parts Specialist A says that the refrigerant leaving a compressor is a high-pressure liquid. Parts Specialist B says that the refrigerant entering the compressor is a low pressure vapor. Who is right?
A. A only
B. B only
C. Both A and B
D. Neither A nor B   (C.12.2)

112. The following are all part of the accessory electrical system EXCEPT:
A. a radio or sound system.
B. an alternator or generator.
C. power seats or windows.
D. a rear window defogger.   (C.13.1)

113. ISO is an abbreviation for:
    A. International Standards Organization.
    B. International Standardized Operations.
    C. Internal Secondary Oscillator.
    D. Internal Scanning Oscilloscope. (C.14.2)

114. Parts Specialist A says that all bolts securing a part should be of the same grade. Parts Specialist B says that a nut should be the same grade as the bolt it is used on. Who is right?
    A. A only
    B. B only
    C. Both A and B
    D. Neither A nor B (C.14.4)

115. Parts Specialist A says the paint topcoat is about 0.04 inch (1.02-mm) thick when applied to metal. Parts Specialist B says the paint topcoat is about 0.06 inch (1.52 mm) thick when applied to plastic. Who is right?
    A. A only
    B. B only
    C. Both A and B
    D. Neither A nor B (C.14.7)

116. Which type fitting will include a ferrule?
    A. Compression
    B. Pipe
    C. Flare
    D. Flange (C.14.6)

117. Hose construction is being discussed, Parts Specialist A says that all hoses used in automotive service must withstand high pressure. Parts Specialist B says that some hoses used in automotive service are made of a reinforced synthetic rubber. Who is right?
    A. A only
    B. B only
    C. Both A and B
    D. Neither A nor B (C.14.8)

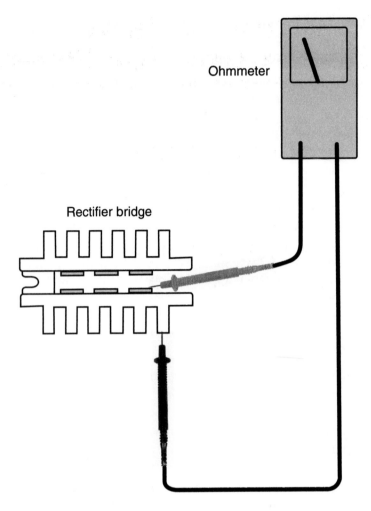

Ohmmeter

Rectifier bridge

118. While discussing the meter setup, Parts Specialist A says the ground of the rectifier is being checked. Parts Specialist B says a diode in the rectifier is being checked. Who is right?

A. A only

B. B only

C. Both A and B

D. Neither A nor B   (C.13.2)

6 Additional Test Questions for Practice

Additional Test Questions

Please note the letter and number in parentheses following each question. They match the overview in section 4 that discusses the relevant subject matter. You may want to refer to the overview using this cross-referencing key to help with questions posing problems for you.

1. Parts Specialist A says a 15 percent discount for $60.00 is $4.00. Parts Specialist B says a part bought for $100.00 and sold for $175.00 will generate a 50 percent profit. Who is right?
   A. A only
   B. B only
   C. Both A and B
   D. Neither A nor B   (A.1)

2. If a customer returns a $250.00 part and is charged a 10 percent restocking fee, how much money is returned to the customer?
   A. $50.00
   B. $100.00
   C. $1000.00
   D. $225.00   (A.2)

3. Parts Specialist A says a 5-liter engine has a displacement of 350 cubic inches. Parts Specialist B says a thermostat marked 192°F is using metric units. Who is right?
   A. A only
   B. B only
   C. Both A and B
   D. Neither A nor B   (A.3)

4. Which of these numbers would appear first in an alphanumeric listing?
   A. 88972E
   B. 89972B
   C. 89973A
   D. 88973A   (A.4)

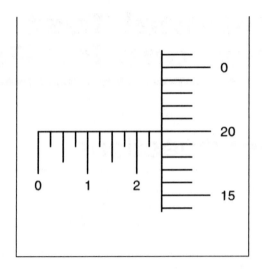

5. The 0–1 inch micrometer reading shown in the illustration is:
   A. 0.220 inch.
   B. 0.225 inch.
   C. 0.245 inch.
   D. 0.250 inch.   (A.5)

6. Parts Specialist A says that a parts specialist should be a problem solver. Parts Specialist B says that a parts specialist should be a person who gains satisfaction through serving and helping others. Who is right?
   A. A only
   B. B only
   C. Both A and B
   D. Neither A nor B   (A.8)

7. Parts Specialist A says that the parts specialist is responsibe for keeping the floors clean. Parts Specialist B says that the parts specialist is only responsible for helping customers. Who is right?
   A. A only
   B. B only
   C. Both A and B
   D. Neither A nor B   (A.9)

8. Parts Specialist A says that you should always lift within your ability. Parts Specialist B says that auto parts are light and you should have no problem lifting them. Who is right?
   A. A only
   B. B only
   C. Both A and B
   D. Neither A and B   (A.11)

9. Parts Specialist A says that any company with hazardous chemicals must maintain Material Safety Data Sheets (MSDSs). Parts Specialist B says that cutting oil is not a chemical that is hazardous. Who is right?
   A. A only
   B. B only
   C. Both A and B
   D. Neither A nor B   (A.12)

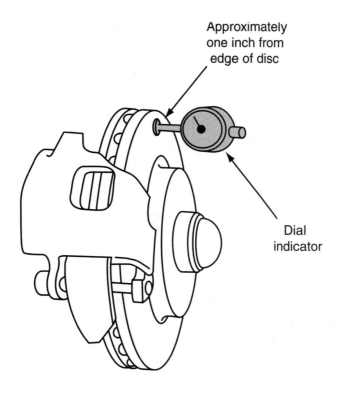

Approximately
one inch from
edge of disc

Dial
indicator

10. Parts Specialist A says the setup is checking the runout of a brake rotor. Parts Specialist B says the setup is measuring the rotor thickness. Who is right?
   A. A only
   B. B only
   C. Both A and B
   D. Neither A nor B   (C.10.4)

11. What kind of merchandise should be displayed by the exit of the store?
   A. Spark plugs
   B. Barrels of oil
   C. Lugnuts
   D. Valve stems   (A.13)

12. Parts Specialist A says that you can assist a do-it-yourselfer without taking too much time by having how-to information at hand. Parts Specialist B says that it is too time consuming to help customers who do not know what they need. Who is right?
   A. A only
   B. B only
   C. Both A and B
   D. Neither A nor B   (B.1)

13. Parts Specialist A says that an opening statement should find out what the customer needs. Parts Specialist B says that an opening statement should put the customer at ease. Who is right?
   A. A only
   B. B only
   C. Both A and B
   D. Neither A nor B   (B.2)

14. Parts Specialist A says to take whatever time is necessary to understand the customer's viewpoint. Parts Specialist B says not to place blame on any one person, but to find a way to correct the problem. Who is right?

    A. A only

    B. B only

    C. Both A and B

    D. Neither A nor B   (B.4)

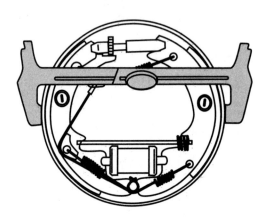

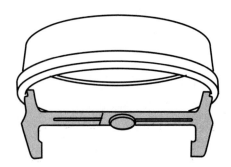

15. While discussing the procedure shown in the figure, Parts Specialist A says the wear of the brake shoes and brake drum is being checked. Parts Specialist B says the procedure is used to set the brake shoes prior to installation. Who is right?

    A. A only

    B. B only

    C. Both A and B

    D. Neither A nor B   (C.10.4)

16. Parts Specialist A says that "good morning" or "good afternoon" is a good opening statement. Parts Specialist B says that you should let the customer start the conversation. Who is right?

    A. A only

    B. B only

    C. Both A and B

    D. Neither A nor B   (B.5)

17. Parts Specialist A says that a counter customer is more important than a telephone customer. Parts Specialist B say that a telephone customer is more important than putting the stock order away. Who is right?

    A. A only

    B. B only

    C. Both A and B

    D. Neither A nor B   (B.6)

18. Parts Specialist A says that the appearance of the store has little influence on the customer. Parts Specialist B says that it is not the parts specialist's responsibility to keep the counter clean. Who is right?

    A. A only

    B. B only

    C. Both A and B

    D. Neither A nor B   (B.8)

19. Parts Specialist A says that if you sell the customers what they need, they will return again and again. Parts Specialist B says that you should promote the sale of items that are on sale. Who is right?

    A. A only

    B. B only

    C. Both A and B

    D. Neither A nor B   (B.9)

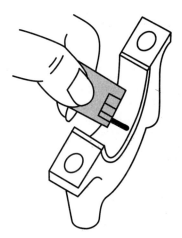

20. Parts Specialist A says the results of this procedure will determine if the bearing needs to be replaced. Parts Specialist B says this procedure will determine if the shaft that revolves on the bearing is out-of-round. Who is right?

    A. A only

    B. B only

    C. Both A and B

    D. Neither A nor B   (C.1.4)

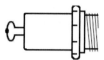

21. What part of the fuel system is shown in the figure?
    A. Fuel filter
    B. Fuel pump
    C. Fuel tank
    D. Fuel pressure regulator   (C.3.1)

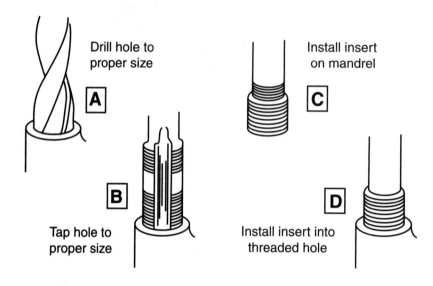

22. Parts Specialist A says the procedure shown is used to repair damaged internal threads. Parts Specialist B says inserts like those being installed in the drawing are available in many different sizes. Who is right?
    A. A only
    B. B only
    C. Both A and B
    D. Neither A nor B   (C.14.2)

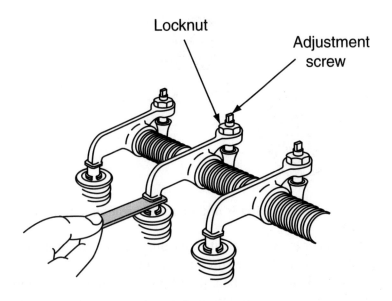

23. Parts Specialist A says the procedure shown will determine if the rocker arms need to be replaced. Parts Specialist B says the procedure adjusts valve lash. Who is right?
    A. A only
    B. B only
    C. Both A and B
    D. Neither A nor B   (C.1.4)

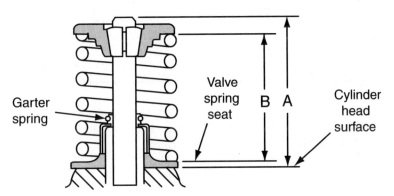

24. Which of the following best describes what dimension B in the drawing represents?
    A. Valve spring free height
    B. Valve spring installed height
    C. Valve stem height
    D. Valve retainer to seat clearance   (C.1.3)

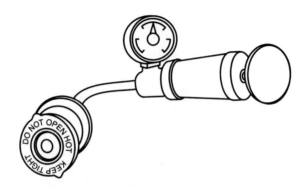

25. Which of the following is not an accurate statement about the use of the tester shown?
    A. The tester is used to create a vacuum on the radiator cap.
    B. The tester creates pressure on the radiator cap.
    C. If the radiator cap is bad, it will not allow pressure to build up.
    D. Every radiator cap is designed to open at a particular pressure.   (C.2.3)

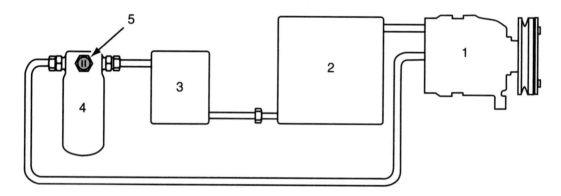

26. While discussing the air conditioning system drawn in the figure, Parts Specialist A says the component labeled 3 is the radiator. Parts Specialist B says the component labeled 1 is the air pump. Who is right?
    A. A only
    B. B only
    C. Both A and B
    D. Neither A nor B   (C.12.1)

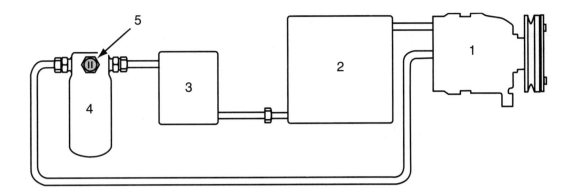

27. Which of the following is NOT a true statement about the components given in the figure?
    A. Component 1 is part of both the low- and high-pressure systems of the A/C unit.
    B. Component 2 changes hot vapor into a cool liquid.
    C. Component 3 changes a cool liquid into a cool vapor.
    D. Component 4 prevents vapor from reaching the compressor.　(C.12.2)

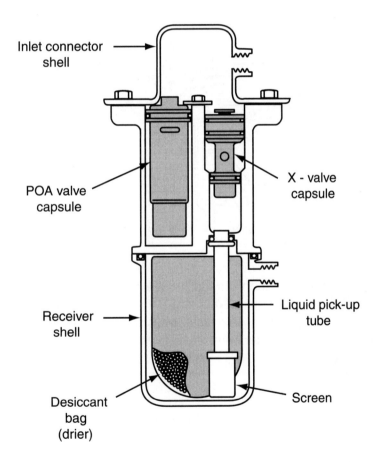

28. While discussing the unit shown, Parts Specialist A says the desiccant bag in this unit is replaceable. Parts Specialist B says this unit controls refrigerant flow through the evaporator and removes moisture from the refrigerant. Who is right?
    A. A only
    B. B only
    C. Both A and B
    D. Neither A nor B　(C.12.2)

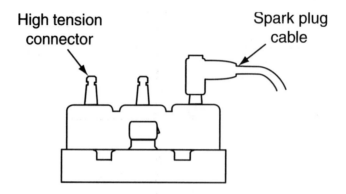

29. The component shown is part of what system?
    A. The ignition system
    B. The fuel injection system
    C. The brake system
    D. The air conditioning system   (C.4.2)

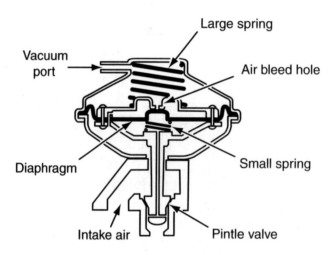

30. Which of the following statements is NOT true?
    A. The valve assembly shown sends a sample of engine exhaust into the cylinders during certain operating conditions.
    B. The valve reduces crankcase pressures and helps prevent oil dilution.
    C. The valve is either controlled by vacuum or by electronics.
    D. The valve can cause a rough idle if it does not fully close.   (C.6.2)

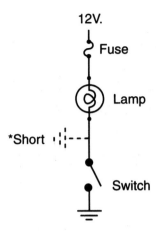

31. While discussing the effects of the short shown in the figure, Parts Specialist A says the lamp would be lit at all times regardless of the position of the switch. Parts Specialist B says the short would cause the fuse to blow. Who is right?

    A. A only

    B. B only

    C. Both A and B

    D. Neither A nor B   (C.13.4)

32. Parts Specialist A says that the thread pitch of an American bolt is measured by the distance between the threads in inches. Parts Specialist B says the thread pitch of a metric bolt is measured by the distance between the threads in millimeters. Who is right?

    A. A only

    B. B only

    C. Both A and B

    D. Neither A nor B   (C.14.1)

33. Part Specialist A says that the vehicle identification number (VIN) is attached to the driver's side of the instrument panel. Parts Specialist B says that the vehicle identification number (VIN) is not visible from the outside of the vehicle. Who is right?

    A. A only

    B. B only

    C. Both A and B

    D. Neither A nor B   (D.1)

34. Parts Specialist A says that the standard location for the production date is on the passenger side of the instrument panel that is visible through the windshield. Parts Specialist B says that the standard location for the production date is in the trunk on the spare tire cover. Who is right?

    A. A only

    B. B only

    C. Both A and B

    D. Neither A nor B   (D.2)

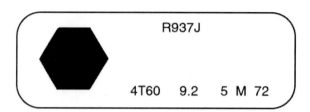

35. Two parts specialists are discussing the identification (ID) tag shown in the figure. Parts Specialist A says that this tag could be from an axle. Parts Specialist B says instead of using a tag, the information could have been stamped into the axle housing. Who is right?
    A. A only
    B. B only
    C. Both A and B
    D. Neither A nor B   (D.3)

36. All of the following are types of cabs EXCEPT a:
    A. crew cab.
    B. long cab.
    C. regular cab.
    D. extended cab.   (D.4)

37. Parts Specialist A says that to find the paint code one can look on the service part identification label. Parts Specialist B says that the service parts identification label can be mounted under the hood or on the driver's door. Who is right?
    A. A only
    B. B only
    C. Both A and B
    D. Neither A nor B   (D.6)

38. Parts Specialist A says that most catalogs do not have dates on them. Parts Specialist B says that all catalogs are the same size so they will fit a standard binder. Who is right?
    A. A only
    B. B only
    C. Both A and B
    D. Neither A nor B   (E.1)

39. A footnote is used for:
    A. locating special parts.
    B. referencing additional information.
    C. checking inventory.
    D. checking the back order list.   (E.2)

40. Correction bulletins are used to:
    A. update the catalog for new parts releases.
    B. update outdated information.
    C. correct inventory problems.
    D. make corrections during catalog maintenance.   (E.3)

Distance Sensor	Vehicle Speed Sensor	VSS
Distributor Ignition	Distributor Ignition	DI
Distributorless Ignition	Electronic Ignition	EI
DLC (Data Link Connector)	Data LInk Connector	DLC
DLI (Distributorless Ignition)	Electronic Ignition	EI
DS (Detonation Sensor)	Knock Sensor	KS
DTC (Diagnostic Trouble Code)	Diagnostic Trouble Code	DTC
DTM (Diagnostic Test Mode)	Diagnostic Test Mode	DTM
Dual Bed	Three Way + Oxidation Catalytic Converter	TWC + OC
Duty Solenoid for Purge Valve	Evaporative Emission Canister Purge Valve	EVAP Canister Purge Valve
E2PROM (Electrically Erasable Programmable Read Only Memory)	Electrically Erasable Programmable Read Only Memory	EEPROM
Early Fuel Evaporation	Early Fuel Evaporation	EFE
EATX (Electronic Automatic Transmission/ Transaxle)	Automatic Transmission	A/T
	Automatic Tranaxle	A/T
EC (Engine Control)	Engine Control	EC
ECA (Electronic Control Assembly)	Powertrain Control Module	PCM
ECL (Engine Coolant Level)	Engine Coolant Level	ECL
ECM (Engine Control Module)	Engine Control Module	ECM
ECT (Engine Coolant Temperature)	Engine Coolant Temperature	ECT
ECT (Engine Coolant Temperature) Sender	Engine Coolant Temperature Sensor	ECT Sensor
ECT (Engine Coolant Temperature) Sensor	Engine Coolant Temperature Sensor	ECT Sensor
ECT (Engine Coolant Temperature) Switch	Engine Coolant Temperature Switch	ECT Switch

41. In the figure, the abbreviation DLC means:

   A. Data Link Connector.

   B. Digital Link Connector.

   C. Diagnostic Link Connector.

   D. Distributorless Link Connector. (E.4)

42. Parts Specialist A says that car batteries are more likely to fail in the winter season so more batteries should be ordered and in stock for the winter season. Parts Specialist B says in the summer months it is not necessary to stock windshield washer fluid because no sales will be made. Who is right?

   A. A only

   B. B only

   C. Both A and B

   D. Neither A nor B (G.4)

43. To keep shelf appearance and condition good, shelves should be:

   A. rebuilt once a year.

   B. painted bright colors.

   C. labeled.

   D. taller than six feet. (G.3)

44. Parts Specialist A says that a parts specialist sometimes will post the cost of the item on a brightly colored poster board to attract the attention of customers. Parts Specialist B says that the display by itself should be enough to let customers know what is on sale. Who is right?

   A. A only

   B. B only

   C. Both A and B

   D. Neither A nor B (G.2)

45. Parts Specialist A says that the parts specialist should build a display with the product being sold because this will be appealing to customers and it will act as an incentive to buy. Parts Specialist B says that the parts specialist should build the display as tall as possible so that it can be seen wherever you are in the store or department. Who is right?

A. A only

B. B only

C. Both A and B

D. Neither A nor B   (G.1)

46. A customer orders six liters of transmission fluid. How many quarts should the parts specialist give to the customer?

A. 2

B. 4

C. 5

D. 6   (A.3)

PORT 2	PARTS-J3	INVOICING	10OCT00 1337

Invoice : 12223   Sale Type: CASH
Customer: 99999
        Name: CASH RETAIL   Parts: 63.20
     Address: Freight: 0.00
 City,St Zip: Tax: 3.79
  Home Phone: Total Invoice: 66.99
                                               Backorder Amount: 0.00

Emp: 117   Sales Person:-   Ship Via: carry   PO: none   B/L:-

Part No.	Description	Bin	O.H.	Cost	Sale	Ext. sale	Q.S.	# A	0.0.	PM
23196550	TIMING BELT	245	11	26.73	37.42	37.42 1				
74073158	BELT, SERP	213	3	11.46	16.04	16.04 1				
21646146	PAPER RAGS	218	8	3.52	4.93	4.93 1				
7148887	DRESSING	13B	5	2.19	3.07	3.07 1				
R23895	TAPE	29H	1	1.24	1.74	1.74 1			M	

F1=Help   F3=Save   F4=Cancel   F8=Print   F10=login

47. According to the invoice, how much did the customer pay for rags?

A. $2.19

B. $3.07

C. $3.52

D. $4.93   (A.6)

48. Parts Specialist A says that charge accounts are for people who do not have the money at the time of their purchase. Parts Specialist B says that a parts specialist must be familiar with the store's charge policies before making a charge transaction. Who is right?

A. A only

B. B only

C. Both A and B

D. Neither A nor B   (A.7)

49. Parts Specialist A says that once a parts specialist has been trained, he or she should be on their own and be capable of performing their job responsibilities without assistance. Parts Specialist B says an experienced parts specialist may be required to assist in training a new employee to become a parts specialist. Who is right?

    A. A only
    B. B only
    C. Both A and B
    D. Neither A nor B   (A.10)

50. Parts Specialist A says an 8 percent (8%) discount for a $20.00 order is $5.00. Parts Specialist B says a part bought for $32.00 and sold for $48.00 generates a 50 percent (50%) profit. Who is right?

    A. A only
    B. B only
    C. Both A and B
    D. Neither A nor B   (A.1)

51. If a customer returns a $50.00 part and is charged a 15 percent (15%) restocking fee, how much money is returned to the customer?

    A. $42.50
    B. $32.50
    C. $22.50
    D. $7.50   (A.2)

52. Which of the following numbers would appear first in an alphanumeric listing?

    A. 10012E
    B. 10002A
    C. 10112B
    D. 10001Z   (A.4)

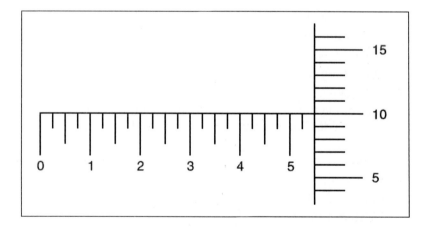

53. Parts Specialist A says the micrometer reading in the figure is 0.560 mm. Parts Specialist B says the micrometer reading in the figure is 0.560 inch. Who is right?

    A. A only
    B. B only
    C. Both A and B
    D. Neither A nor B   (A.5)

54. Parts Specialist A says that parts specialists have to work with customers, suppliers, manufacturers, and other people in the aftermarket industry. Parts Specialist B says that parts specialists only have to work with the customers that come to the counter. Who is right?
    A. A only
    B. B only
    C. Both A and B
    D. Neither A nor B   (A.8)

55. A parts specialist is responsible for keeping all of the following items clean and orderly EXCEPT:
    A. floors.
    B. shelves.
    C. shop parts cleaner.
    D. displays.   (A.9)

56. Parts Specialist A says that small parts usually do not weigh much, therefore there is no need to seek help when handling them. Parts Specialist B says that knowing the proper way to lift and work within one's ability is important for handling parts. Who is right?
    A. A only
    B. B only
    C. Both A and B
    D. Neither A nor B   (A.11)

57. A company that uses hazardous chemicals must follow the guidelines of the:
    A. Environmental Protection Agency (EPA).
    B. Occupational Safety and Health Administration (OSHA).
    C. Department of Public Information Office (DPIO).
    D. Material Safety Data Sheets (MSDS).   (A.12)

58. What should be the LEAST likely item found near a store exit?
    A. Barrels of oil
    B. 10-pound bags of floor sweep
    C. 6-gallon gasoline cans
    D. Spark plug display   (A.13)

59. Parts Specialist A says that a parts specialist should always sell the customers what they want even if it will not fix the problem. Parts Specialist B says that only parts that are needed for a particular repair should be sold to the customers. Who is right?
    A. A only
    B. B only
    C. Both A and B
    D. Neither A nor B   (B.1)

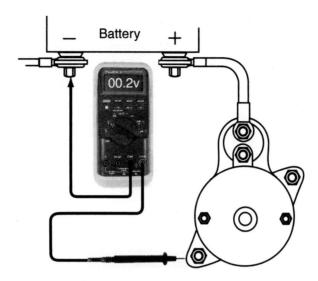

60. Which of the following describes what the meter setup is measuring?
    A. Voltage drop across the ground circuit
    B. Available voltage at the starter
    C. Voltage drop of the starter motor
    D. Voltage at the battery   (C.13.4)

61. Parts Specialist A says that "How can I help you?" is the best opening question
    that may be made to a customer. Parts Specialist B says that "May I help you?" is
    one of the worst greetings a parts specialist could make to a customer. Who is
    right?
    A. A only
    B. B only
    C. Both A and B
    D. Neither A nor B   (B.2)

62. Parts Specialist A says that an embarrassed customer will probably spend their
    money elsewhere. Parts Specialist B says that the only way to resolve a problem is
    to find out who is to blame. Who is right?
    A. A only
    B. B only
    C. Both A and B
    D. Neither A nor B   (B.4)

63. Parts Specialist A says that eye contact should be made with each customer as they
    enter the store. Parts Specialist B says that eye contact should be maintained with
    the customer while looking in a parts catalog. Who is right?
    A. A only
    B. B only
    C. Both A and B
    D. Neither A nor B   (B.5)

64. Parts Specialist A says that a telephone customer cannot see the parts specialist's body language so all communication must be accomplished with words, tones, timing, and inflection. Parts Specialist B says that a parts specialist must remember to speak clearly and slowly enough to be understood and must request that the customer do so if he or she cannot be understood. Who is right?

   A. A only
   B. B only
   C. Both A and B
   D. Neither A nor B   (B.6)

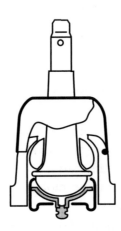

65. While discussing the suspension part shown in the drawing, Parts Specialist A says it is part of the steering linkage and is called a tie rod end. Parts Specialist B says the part allows the wheel spindle to rotate when the steering wheel is turned. Who is right?

   A. A only
   B. B only
   C. Both A and B
   D. Neither A nor B   (C.11.2)

66. Parts Specialist A says that it is important to keep a rag handy for wiping dirt and grime from the counter. Parts Specialist B says that the parts specialist is in charge of the store layout. Who is right?

   A. A only
   B. B only
   C. Both A and B
   D. Neither A nor B   (B.8)

67. Parts Specialist A says that a fuel filter is used to remove dirt from the fuel. Parts Specialist B says that a vehicle could have a carburetor or fuel injection. Who is right?

   A. A only
   B. B only
   C. Both A and B
   D. Neither A nor B   (C.3.2)

68. Parts Specialist A says that depending on the electric ignition system, the coils may be serviced separately or as a complete unit. Parts Specialist B says that the computer, ignition module, and position sensor combine to control spark timing and advance. Who is right?
   A. A only
   B. B only
   C. Both A and B
   D. Neither A nor B   (C.4.2)

69. Parts Specialist A says that the exhaust system helps to heat the vehicle. Parts Specialist B says that the exhaust manifold may be made of thermoplastic. Who is right?
   A. A only
   B. B only
   C. Both A and B
   D. Neither A nor B   (C.5.2)

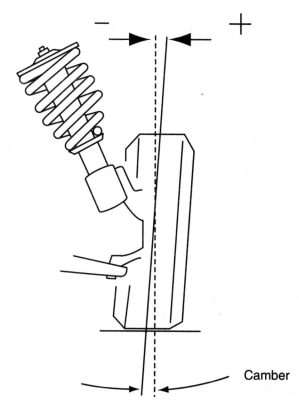

Camber

70. Which of the following statements about the alignment angle shown is NOT correct?
   A. Camber changes as a vehicle moves down the road.
   B. Incorrect camber will cause excessive tire wear.
   C. Camber is adjusted by changing the length of the tie rods.
   D. Camber can be affected by a variety of damaged suspension parts.   (C.11.2)

71. Parts Specialist A says that on many new models, emissions systems are not used. Parts Specialist B says that the EGR valve is controlled by a spring and linkage. Who is right?
   A. A only
   B. B only
   C. Both A and B
   D. Neither A nor B   (C.6.2)

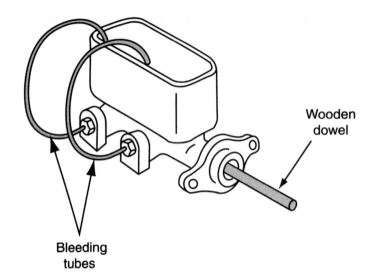

72. Which of the following statements about the procedure shown in the figure is NOT true?
    A. This should be done before installing a new brake master cylinder.
    B. This is done to remove air from the internal pistons in the cylinder.
    C. The cylinder must have fluid in it prior to performing this procedure.
    D. The dowel and tubes must be kept in place during installation.   (C.10.4)

73. Parts Specialist A says that front-wheel-drive vehicles have a driveshaft for each front wheel. Parts Specialist B says that the outboard constant velocity (CV) joint is a plunger type. Who is right?
    A. A only
    B. B only
    C. Both A and B
    D. Neither A nor B   (C.9.2)

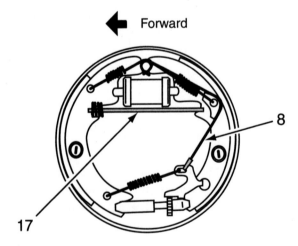

74. Parts Specialist A says if part 8 in the figure is broken or damaged, the brakes will function normally but will not self-adjust. Parts Specialist B says part 17 helps keep the shoes somewhat expanded toward the drum. Who is right?
    A. A only
    B. B only
    C. Both A and B
    D. Neither A nor B   (C.10.4)

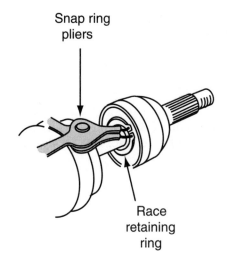

Snap ring pliers

Race retaining ring

75. Parts Specialist A says the part being removed in the drawing is an outboard CV joint. Parts Specialist B says the retaining ring on the part shown should not be reused. Who is right?

A. A only

B. B only

C. Both A and B

D. Neither A nor B   (C.9.4)

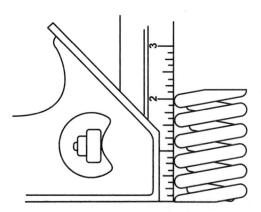

76. While discussing the setup shown in the figure, Parts Specialist A says if the spring is distorted, it should be compressed and heated until it is square. Parts Specialist B says if the freestanding height of the spring is lower than specs, the problem should be corrected by the use of shims installed at the spring seat. Who is right?

A. A only

B. B only

C. Both A and B

D. Neither A nor B   (C.1.4)

77. Parts Specialist A says that the vehicle's electrical system is protected by fuses, circuit breakers, and fusible links. Parts Specialist B says that a shorted circuit may blow a fuse or open a circuit breaker or fusible link. Who is right?

A. A only

B. B only

C. Both A and B

D. Neither A nor B   (C.13.2)

78. A bolt with three radial lines embossed on the head would be:
    A. grade 1.
    B. grade 3.
    C. grade 5.
    D. metric.   (C.14.1)

79. Parts Specialist A says that the identification (ID) data could be stamped into the casting. Parts Specialist B says that the identification (ID) data is on all of the major components of a vehicle. Who is right?
    A. A only
    B. B only
    C. Both A and B
    D. Neither A nor B   (D.3)

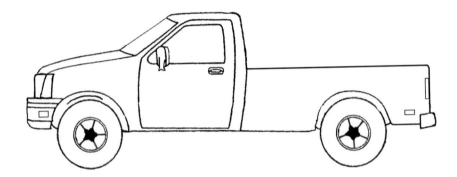

80. Two parts specialists are discussing the vehicle shown in the figure. Parts Specialist A says that the vehicle in the illustration is a crew cab vehicle. Parts Specialist B says that an extended cab vehicle could have two, three, or four doors. Who is right?
    A. A only
    B. B only
    C. Both A and B
    D. Neither A nor B   (D.4)

81. Most catalogs contain all of the following information EXCEPT:
    A. form number(s).
    B. date of issue.
    C. the name of product lines and manufacturers.
    D. correction form(s).   (E.1)

82. What information would most likely be found in a footnote?
    A. Alternative part information
    B. Back order list
    C. Shipping size
    D. Nearest availability   (E.2)

83. Parts Specialist A says that supersession bulletins may include part numbers that supersede previously issued part numbers. Parts Specialist B says that technical bulletins may alert counter personnel to any unusual installation or application problems. Who is right?
    A. A only
    B. B only
    C. Both A and B
    D. Neither A nor B   (E.3)

84. Parts Specialist A says that to verify an abbreviation one should refer to the abbreviation list. Parts Specialist B says that abbreviations are only used in footnotes. Who is right?
    A. A only
    B. B only
    C. Both A and B
    D. Neither A nor B   (G.4)

85. Which of the following items is LEAST likely to be sold in the summer?
    A. Air conditioning refrigerant
    B. A car battery
    C. Dry gas
    D. Engine coolant   (G.4)

86. All of the following statements are true about shelf appearance and condition EXCEPT:
    A. Items on a shelf should be clean and new looking.
    B. Returned items cannot be put back on a shelf because they will look used.
    C. Items should be pulled forward to give a full and easy-to-find appearance.
    D. Labels can be used to help organization.   (G.3)

87. Which of the following display pricing methods is the LEAST effective method used to promote a product?
    A. Tagging items on the shelf with a price
    B. Setting up a display of the product with an explanation and the price
    C. Displaying prices in the window for potential customers to see
    D. Distributing fliers to encourage product sales   (G.2)

88. The LEAST important part of a display is the:
    A. location of the display.
    B. arrangement of the display.
    C. product on the display.
    D. cost of the item on the display.   (G.1)

89. Which type of fitting relies on tapered threads to prevent leaks?
    A. Pipe
    B. Compression
    C. Flare
    D. Tubing   (C.14.3)

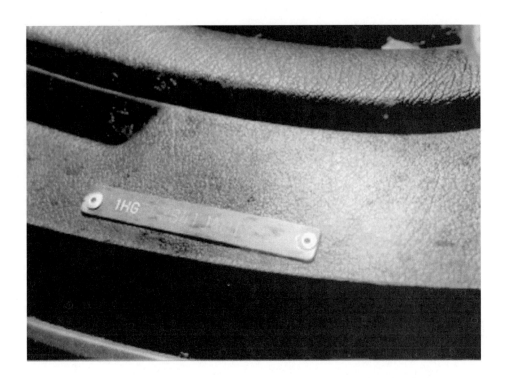

90. Two parts specialists are looking at the vehicle identification number (VIN) in the figure. Parts Specialist A says that the VIN is visible through the windshield. Parts Specialist B says that the VIN is located on the driver's side of the instrument panel. Who is right?

    A. A only

    B. B only

    C. Both A and B

    D. Neither A nor B   (D.1)

91. An impulse item is:

    A. something a customer needs to fix his or her vehicle.

    B. something a vendor will stock if it can be purchased cheaply.

    C. a flashy item that may catch a customer's eye while browsing and that may lead to a potential sale.

    D. an item needed to make an emergency repair.   (G.4)

92. While discussing shelf appearance conditions, Parts Specialist A says that rotating stock on the shelves keeps items from expiring or becoming old-looking or dusty. Parts Specialist B says that nothing in a parts store will expire and to simply pull all the items off the shelves, dust, and replace them. Who is right?

    A. A only

    B. B only

    C. Both A and B

    D. Neither A nor B   (G.3)

93. A customer orders three liters of oil. How many quarts would the parts specialist give the customer?

    A. 2

    B. 4

    C. 3

    D. 6   (A.3)

94. Parts Specialist A says that 327 cubic inches is equal to about 5.8 liters. Parts Specialist B says that 488 cubic inches is equal to 8 liters. Who is right?
    A. A only
    B. B only
    C. Both A and B
    D. Neither A nor B   (A.3)

PORT 2   PARTS-J3   INVOICING   10OCT00 1337

Invoice : 12223   Sale Type: CASH
Customer: 99999
    Name: CASH RETAIL   Parts: 165.94
 Address: Freight: 0.00
City,St Zip: Tax: 9.96
Home Phone: Total Invoice: 175.90
                                                Backorder Amount: 0.00

Emp: 117   Sales Person:-   Ship Via: carry   PO: none   B/L:-

Part No.	Description	Bin	O.H.	Cost	Sale	Ext. sale	Q.S.	# A	0.0.	PM
23196550	CAMSHAFT	245	11	89.18	124.85	124.85	1			
74073158	NUT	213	3	0.34	0.48	0.95	2	M		
21646146	PAPER RAGS	218	8	3.52	4.93	4.93	1			
7148887	TIMING BELT	13B	5	22.54	31.56	31.56	1			
R23895	OIL SEAL	29H	22	2.61	3.65	3.65	1			

F1=Help   F3=Save   F4=Cancel   F8=Print   F10=login

95. While discussing the information on the invoice, Parts Specialist A says the markup on the nut is sixty-one cents. Parts Specialist B says the markup on the camshaft is 40 percent. Who is right?
    A. A only
    B. B only
    C. Both A and B
    D. Neither A nor B   (A.6)

96. Parts Specialist A says that charge accounts are commonly used for large sales. Parts Specialist B says that on some charge accounts the customer may not have to pay sales tax. Who is right?
    A. A only
    B. B only
    C. Both A and B
    D. Neither A nor B   (A.7)

97. Parts Specialist A says it is better to ask for help than to sell the customer the wrong parts. Parts Specialist B says that customer satisfaction is built by supplying the right information and the correct parts. Who is right?
    A. A only
    B. B only
    C. Both A and B
    D. Neither A nor B   (A.10)

98. Parts Specialist A says that a purchase order is used to purchase parts for the store. Parts Specialist B says to only make an emergency order if the vehicle does not have to be worked on right away. Who is right?

    A. A only

    B. B only

    C. Both A and B

    D. Neither A nor B   (A.14)

99. The following are all types of orders EXCEPT:

    A. stock order.

    B. back order.

    C. emergency order.

    D. purchase order.   (A.14)

100. Parts Specialist A says that a lot of the time customers simply lack confidence rather than skill. Parts Specialist B says that most vehicle problems are beyond the do-it-yourselfer's capabilities. Who is right?

    A. A only

    B. B only

    C. Both A and B

    D. Neither A nor B   (B.3)

101. Parts Specialist A says that confidence can be boosted by providing information about the repair. Parts Specialist B says that parts specialists should always avoid talking down to the customer. Who is right?

    A. A only

    B. B only

    C. Both A and B

    D. Neither A nor B   (B.3)

102. Defensive selling is a method of selling that will:

    A. increase the amount of sales for a store.

    B. increase the amount of customer returns.

    C. make shopping easier for the customer.

    D. protect the interests of the store.   (B.7)

103. Parts Specialist A says that some customers are just parts replacers; when they buy a part that does not repair the vehicle, they try to return or exchange the part for one that will fix the problem. Parts Specialist B says that defensive selling is one way to keep customers from returning parts that have been installed. Who is right?

    A. A only

    B. B only

    C. Both A and B

    D. Neither A nor B   (B.7)

104. Selling related parts could increase profits by what percentage?

    A. 80 percent

    B. 100 percent

    C. 30 percent

    D. 5 percent   (B.9)

105. Parts Specialist A says to only sell parts that the customer asks for. Parts Specialist B says to only suggest related items or parts if questioned by the customer. Who is right?
    A. A only
    B. B only
    C. Both A and B
    D. Neither A nor B   (B.9)

106. Parts Specialist A says that it is the parts specialist's job to sell the most expensive part every time. Parts Specialist B says that the parts specialist should tell the customer what they need. Who is right?
    A. A only
    B. B only
    C. Both A and B
    D. Neither A nor B   (B.10)

107. Two parts specialists are discussing how to sell brake shoes to a customer. There are four different variations of brake shoes in stock that fit the customer's vehicle. Parts Specialist A says that no matter how many different kinds of brake shoes the store carries, it is still the customer's decision. Parts Specialist B says that a parts specialist should always explain the benefits and cost of each kind of brake shoe. Who is right?
    A. A only
    B. B only
    C. Both A and B
    D. Neither A nor B   (B.10)

108. Parts Specialist A says that a parts specialist should always close the sale. Parts Specialist B says that customers come into the store only to buy. Who is right?
    A. A only
    B. B only
    C. Both A and B
    D. Neither A nor B   (B.11)

109. Parts Specialist A says that after some years of experience, a parts specialist will be able to identify the different types of customers. Parts Specialist B says that some customers come into the store just to look around or to compare prices. Who is right?
    A. A only
    B. B only
    C. Both A and B
    D. Neither A nor B   (B.11)

110. A parts specialist is serving a customer at the counter and the phone rings. The parts specialist must put the telephone customer on hold. How should the parts specialist handle this situation?
    A. Answer the telephone, identify the business, and politely say, "Please hold one moment."
    B. Just push the hold button.
    C. Answer the telephone and serve the phone customer while the counter customer waits.
    D. Yell for someone to pick up the telephone.   (B.12)

111. Two parts specialists are discussing how to handle a customer at the counter while the telephone is ringing. Parts Specialist A says to let the phone ring and hope someone else answers it. Parts Specialist B says to answer the phone and ask the calling customer to hold. Who is right?
    A. A only
    B. B only
    C. Both A and B
    D. Neither A nor B   (B.12)

112. Parts Specialist A says that one should promote sales and service whenever possible. Parts Specialist B says that machine shop service is one area where sales can be promoted. Who is right?
    A. A only
    B. B only
    C. Both A and B
    D. Neither A nor B   (B.13)

113. Parts Specialist A says that it is not necessary to promote sales. Parts Specialist B says that most customers will know if you have a machine shop. Who is right?
    A. A only
    B. B only
    C. Both A and B
    D. Neither A nor B   (B.13)

114. Parts Specialist A says that displays should be set up to promote related sales. Parts Specialist B says that most manufacturers are putting parts in colorful boxes to help promote sales when used in a display. Who is right?
    A. A only
    B. B only
    C. Both A and B
    D. Neither A nor B   (B.14)

115. Parts Specialist A says that making related sales helps build profits. Parts Specialist B says that making related sales helps the customer as well as the store. Who is right?
    A. A only
    B. B only
    C. Both A and B
    D. Neither A nor B   (B.14)

116. Parts Specialist A says that the goal is to give the customer the right part every time. Parts Specialist B says that if the customer does not offer enough information to obtain the correct part, it is acceptable to make an educated guess. Who is right?
    A. A only
    B. B only
    C. Both A and B
    D. Neither A nor B   (B.15)

117. Parts Specialist A says not to give any extra information about a given system, even when the parts specialist knows that the customer is buying the wrong part. Parts Specialist B says to give suggestions on how to fix the vehicle whenever possible. Who is right?
    A. A only
    B. B only
    C. Both A and B
    D. Neither A nor B   (B.15)

118. Parts Specialist A says that a good sale closer always lets the customer know that they expect the sale. Parts Specialist B says that it is not part of a parts specialist's job to close the sale. Who is right?
   A. A only
   B. B only
   C. Both A and B
   D. Neither A nor B (B.16)

119. Parts Specialist A says that the closing is complete when the parts specialist has a commitment from the customer. Parts Specialist B says that confidence in one's selling ability will help in the store's sales. Who is right?
   A. A only
   B. B only
   C. Both A and B
   D. Neither A nor B (B.16)

120. Parts Specialist A says that the standard location for the production date is on the underside of the trunk lid. Parts Specialist B says that the production date is used to tell the parts specialist when service was last completed on the vehicle. Who is right?
   A. A only
   B. B only
   C. Both A and B
   D. Neither A nor B (D.2)

121. Parts Specialist A says that the vehicle build sheet includes vehicle option content information. Parts Specialist B says that not all vehicles have a build sheet. Who is right?
   A. A only
   B. B only
   C. Both A and B
   D. Neither A nor B (D.5)

122. Where on the vehicle would the axle ratio information be found?
   A. Emission label under the hood
   B. Vehicle build sheet
   C. Vehicle identification number (VIN)
   D. Axle tag (D.5)

123. Parts Specialist A says that the paint code is on the inside of the front bumper. Parts Specialist B says that most vehicles do not have a paint code anywhere on the vehicle. Who is right?
   A. A only
   B. B only
   C. Both A and B
   D. Neither A nor B (D.6)

124. Parts Specialist A says that product additions can be found in the table of contents of any parts catalog. Parts Specialist B says that, in most cases, the index will not help in finding a part in a parts catalog. Who is right?
   A. A only
   B. B only
   C. Both A and B
   D. Neither A nor B (E.5)

125. Parts Specialist A says that an index will not indicate if the catalog is for passenger cars or trucks. Parts Specialist B says that parts illustrations can be found by using the table of contents. Who is right?
    A. A only
    B. B only
    C. Both A and B
    D. Neither A nor B   (E.5)

126. Parts Specialist A says that technical letters help to keep the catalogs updated. Parts Specialist B says that it is the job of the parts delivery driver to update the parts catalogs. Who is right?
    A. A only
    B. B only
    C. Both A and B
    D. Neither A nor B   (E.6)

127. Parts Specialist A says that it is very important that parts specialists use the most up-to-date information. Parts Specialist B says that there is no way to keep the catalogs updated unless you buy new catalogs. Who is right?
    A. A only
    B. B only
    C. Both A and B
    D. Neither A nor B   (E.6)

128. What information does a packing list contain?
    A. The type of vehicle the delivery person is driving
    B. Where the delivery person will be delivering the shipment
    C. The selling store's business hours
    D. A parts list and how many items are in the shipment   (F.2)

129. Parts Specialist A says that a packing list should be included with each shipment. Parts Specialist B says that one packing list can contain more than one part. Who is right?
    A. A only
    B. B only
    C. Both A and B
    D. Neither A nor B   (F.2)

130. The definition of physical inventory is:
    A. seeing what the computers have recorded as your inventory.
    B. looking at what was sold for the last year to determine what should be stocked for the next year.
    C. pulling all the stock off of the shelves and counting what is in the inventory and comparing the count to what the computer indicates for your inventory.
    D. looking for missing inventory in the back room of the store.   (F.3)

131. Parts Specialist A says that a physical inventory should be done once a year. Parts Specialist B says that keeping good inventory will help ensure that parts are in stock for the customer when needed. Who is right?
    A. A only
    B. B only
    C. Both A and B
    D. Neither A nor B   (F.3)

132. After the physical inventory has been completed, the parts specialist should:
    A. fill out a back order form.
    B. reload the computer software.
    C. report all discrepancies.
    D. rotate the stock.   (F.4)

133. Parts Specialist A says that all of the discrepancy forms should be kept and filed. Parts Specialist B says that after a discrepancy form has been filled out, it must be corrected in the computer system. Who is right?
    A. A only
    B. B only
    C. Both A and B
    D. Neither A nor B   (F.4)

134. Stock should be rotated:
    A. once a year.
    B. every time the shelves are stocked.
    C. every month.
    D. weekly.   (F.5)

135. Stock rotation prevents:
    A. older stock from sitting at the back of the shelf and eventually having an aged look.
    B. errors in the inventory.
    C. having to do physical inventory.
    D. having to look at the bulletins before doing catalog maintenance.   (F.5)

136. Parts Specialist A says that a special order is the weekly stock order. Parts Specialist B says that a part that repeatedly appears on the special order list could possibly become a part that is stocked in the store. Who is right?
    A. A only
    B. B only
    C. Both A and B
    D. Neither A nor B   (F.6)

137. Parts Specialist A says that a special order is made for a part that is not stocked in that store. Parts Specialist B says that some parts will always be special order parts. Who is right?
    A. A only
    B. B only
    C. Both A and B
    D. Neither A nor B   (F.6)

138. Parts Specialist A says that a parts specialist must be careful when checking core returns to make sure that they are in rebuildable condition. Parts Specialist B says that only rebuildable parts have core charges. Who is right?
    A. A only
    B. B only
    C. Both A and B
    D. Neither A nor B   (F.7)

139. All of the following parts may have a core charge EXCEPT:
    A. brake shoes.
    B. brake calipers.
    C. water pumps.
    D. spark plugs.   (F.7)

140. A warranty-return part should be:
    A. stored in the same room as good parts.
    B. sent to other stores to show the poor quality.
    C. be kept in the delivery truck for disposal.
    D. kept off the shelf, not to be sold again.   (F.8)

141. Parts Specialist A says that the part remanufacturing company has a form to fill
    out so they know what is wrong with the core that was returned. Parts Specialist B
    says that a brake shoe is not a part that would have a core charge. Who is right?
    A. A only
    B. B only
    C. Both A and B
    D. Neither A nor B   (F.8)

142. A pair refers to:
    A. 1.
    B. 2.
    C. 4.
    D. 6.   (F.9)

143. Each refers to:
    A. 10.
    B. 5.
    C. 2.
    D. 1.   (F.9)

144. An additional freight charge could be added to a:
    A. special order.
    B. stock order.
    C. emergency order.
    D. purchase order.   (F.10)

145. Parts Specialist A says that a restocking charge could be placed on an item that is
    returned for a refund. Parts Specialist B says that freight charges are only added to
    stock orders. Who is right?
    A. A only
    B. B only
    C. Both A and B
    D. Neither A nor B   (F.10)

146. The main benefit of using a lost sales report is:
    A. to keep track of the parts that were stolen by shoplifters.
    B. to keep track of the parts that were sold to professional shops.
    C. to keep track of parts that are selling and those that are not selling.
    D. to keep track of parts that were sold to do-it-yourselfers.   (F.1)

LEE'S AUTOMOTIVE
www.leesezparts.com

Date: ___ / ____ / _____

Customer: _____

Customer Type: _____ **Account #:** _____

Counterperson: _____

Part Manufacturer: _____

Part Number: _____

Reason for Lost Sale: _____

_____

_____

147. What type of report is shown in the figure?
  A. A stock order
  B. A shoplifting report
  C. A special order
  D. A lost sales report   (F.1)

148. Worn valve lock grooves are being discussed. Parts Specialist A says worn valve lock grooves may cause the valve locks to fly out of place with the engine running, resulting in severe engine damage. Parts Specialist B says worn valve lock grooves may cause a clicking noise with the engine idling. Who is right?
  A. A only
  B. B only
  C. Both A and B
  D. Neither A nor B   (C.1.3)

149. A cylinder balance test is being performed on an engine having electronic fuel injection. Cylinder number three provides very little drop in rpm. Parts Specialist A says the ignition system may be misfiring on the number three cylinder. Parts Specialist B says the engine may have an intake manifold vacuum leak. Who is right?
  A. A only
  B. B only
  C. Both A and B
  D. Neither A nor B   (C.1.4)

150. A collapsed upper radiator hose is being discussed. Parts Specialist A says this condition may be caused by a defective pressure release valve in the radiator cap. Parts Specialist B says this condition may be caused by a restriction in the hose between the radiator filler neck and the recovery reservoir. Who is right?
  A. A only
  B. B only
  C. Both A and B
  D. Neither A nor B   (C.2.3)

151. An excessively high coolant level in the recovery reservoir may be caused by any of the following problems EXCEPT:
   A. restricted radiator tubes.
   B. a thermostat stuck open.
   C. a loose water pump impeller.
   D. an inoperative electric-drive cooling fan.   (C.2.4)

152. Reduced turbocharger boost pressure may be caused by:
   A. a wastegate valve stuck closed.
   B. a wastegate valve stuck open.
   C. a leaking wastegate diaphragm.
   D. a disconnected wastegate linkage.   (C.3.3)

153. All of the following statements about fuel pump testing on a port fuel injected engine are true EXCEPT:
   A. The fuel system should be depressurized prior to the fuel pump pressure test.
   B. If the fuel pump pressure is less than specified, the air/fuel ratio is too rich.
   C. The pressure gauge should be connected to the Schrader valve on the fuel rail.
   D. Lower than specified fuel pressure may be caused by a restricted fuel filter.   (C.3.4)

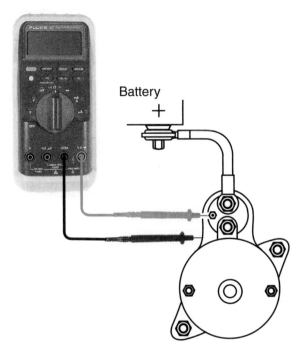

154. Parts Specialist A says the meter hookup shown in the drawing is measuring the available voltage to the starter solenoid. Parts Specialist B says the reading of 12 volts with the setup shown means the solenoid is using all of the available voltage. Who is right?
   A. A only
   B. B only
   C. Both A and B
   D. Neither A nor B   (C.13.4)

155. A low maximum secondary coil voltage may be caused by:
    A. a low primary resistance.
    B. a low primary input voltage.
    C. wide spark plug gaps.
    D. an open spark plug wire. (C.4.4)

156. The topic for discussion is catalytic converters. Parts Specialist A says a gauge reading of 2.0 psi (13.8 kPa) taken during a back pressure test indicates a restricted exhaust system. Parts Specialist B says if the hydrocarbon (HC) level is 200 parts-per-million (ppm) during a temperature test, the catalytic converter is defective. Who is right?
    A. A only
    B. B only
    C. Both A and B
    D. Neither A nor B (C.5.3)

157. Parts Specialist A says the positive crankcase ventilation (PCV) valve moves toward the closed position when the throttle is opened from idle to half throttle. Parts Specialist B says the positive crankcase ventilation (PCV) valve is moved toward the closed position by spring tension. Who is right?
    A. A only
    B. B only
    C. Both A and B
    D. Neither A nor B (C.5.4)

158. The inside of the air cleaner is contaminated with engine oil. Parts Specialist A says the positive crankcase ventilation (PCV) valve clean air filter in the air cleaner may be plugged. Parts Specialist B says the hose from the positive crankcase ventilation (PCV) valve to the intake manifold may be severely restricted. Who is right?
    A. A only
    B. B only
    C. Both A and B
    D. Neither A nor B (C.6.3)

159. When diagnosing a catalytic converter with a digital pyrometer, if the converter is operating properly:
    A. the converter outlet should be 100°F (38°C) hotter than the inlet.
    B. the converter inlet and outlet should be the same temperature.
    C. the converter inlet should be 50°F (10°C) hotter than the outlet.
    D. the converter outlet should be 50°F (10°C) hotter than the inlet. (C.6.4)

160. Repeated extension housing seal failure may be caused by:
    A. a scored driveshaft yoke.
    B. excessive output shaft end play.
    C. excessive input shaft end play.
    D. a worn output shaft bearing. (C.7.3)

161. The second speed gear dog teeth and blocking ring teeth are badly worn. This problem may cause:
    A. a growling noise while driving in second gear.
    B. a vibration while accelerating in second gear.
    C. hard shifting in second and third gear.
    D. the transaxle to jump out of second gear. (C.7.4)

162. The fluid in an automatic transaxle is a dark brown color and has a burned smell. Parts Specialist A says this problem may be caused by a worn front planetary sungear. Parts Specialist B says this problem may be caused by worn friction-type clutch plates. Who is right?
   A. A only
   B. B only
   C. Both A and B
   D. Neither A nor B   (C.8.3)

163. An improper band adjustment may cause:
   A. shifts at a lower speed than specified.
   B. transmission slipping in some gears.
   C. shifts at a higher speed than specified.
   D. transmission slipping in all gears.   (C.8.4)

164. While discussing driveshaft and universal joint (U-joint) diagnosis in rear-wheel-drive (RWD) vehicles, Parts Specialist A says a worn U-joint may cause a squeaking noise that decreases in relation to vehicle speed. Parts Specialist B says a heavy vibration that only occurs during acceleration may be caused by a worn centering ball and socket on a double Cardan U-joint. Who is right?
   A. A only
   B. B only
   C. Both A and B
   D. Neither A nor B   (C.9.3)

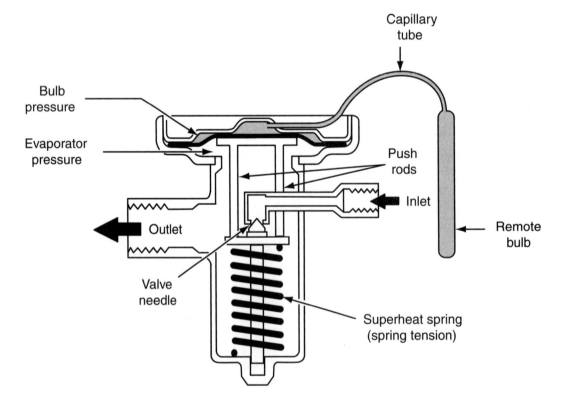

165. The air conditioning part shown in the figure is a(n):
   A. accumulator.
   B. thermal expansion valve.
   C. receiver/drier.
   D. low-pressure relief valve.   (C.12.2)

166. Brake hose replacement is being discussed. Parts Specialist A says that the sealing washer on the male end of a brake hose may be reused. Parts Specialist B says the brake line female fitting end should be installed and tightened before the male end is installed. Who is right?
    A. A only
    B. B only
    C. Both A and B
    D. Neither A nor B   (C.10.3)

167. Parts Specialist A says a bent backing plate may cause brake grabbing. Parts Specialist B says a loose anchor bolt may cause brake chatter. Who is right?
    A. A only
    B. B only
    C. Both A and B
    D. Neither A nor B   (C.10.4)

168. After servicing and lubricating the rear wheel bearings in a front-wheel-drive car, the bearing adjusting nut is tightened to 20 ft.-lbs. (27 N-m) and loosened one-half turn. The next step in the bearing adjusting procedure is to:
    A. install and align the nut retainer and cotter key on the axle.
    B. tighten the adjusting nut so the cotter key hole lines up properly.
    C. tighten the adjusting nut to 10 to 15 ft.-lbs. (13.6 to 20.3 N-m).
    D. loosen the adjusting nut so the cotter key hole lines up properly.   (C.11.3)

169. All of the following statements about caster adjustment are true EXCEPT:
    A. the caster angle is measured with the front wheels straight ahead.
    B. the front wheels are turned 20 degrees outward and then 20 degrees inward to read the caster angle.
    C. the brakes must be applied with a brake pedal jack before reading the caster angle.
    D. the front suspension should be jounced several times before reading the caster angle.   (C.11.4)

170. A receiver/dryer is located between the condenser outlet and the evaporator inlet. Parts Specialist A says the receiver/dryer should be changed if the outlet is colder than the inlet. Parts Specialist B says the receiver/dryer should be changed if the refrigerant in the sight glass appears red. Who is right?
    A. A only
    B. B only
    C. Both A and B
    D. Neither A nor B   (C.12.3)

171. The discharge air temperature is higher than specified in an air conditioning (A/C) heater system with a coolant flow control valve when the system is operating in the maximum A/C mode. Parts Specialist A says the coolant flow control valve may be stuck in the open position. Parts Specialist B says the coolant flow control valve vacuum hose may be disconnected or have a leak. Who is right?
    A. A only
    B. B only
    C. Both A and B
    D. Neither A nor B   (C12.4)

172. While discussing resistance measurement with an ohmmeter, Parts Specialist A says that an ohmmeter may be connected to a circuit in which current is flowing. Parts Specialist B says when testing a spark plug wire with a 20,000 Ohm resistance, use the ×100 meter scale. Who is right?
    A. A only
    B. B only
    C. Both A and B
    D. Neither A nor B   (C.13.3)

173. A wiper circuit has an open shunt coil. Parts Specialist A says this problem may cause the wiper motor to operate only at low speed. Parts Specialist B says this problem may cause the wiper motor to park with the wipers up on the windshield. Who is right?
    A. A only
    B. B only
    C. Both A and B
    D. Neither A nor B   (C.13.4)

174. All of the following statements relating to motor oil are true EXCEPT:
    A. Customers are very sensitive to motor oil prices.
    B. Motor oil is a high volume, high profit item.
    C. Changing motor oil is the most popular DIY product.
    D. Stocking motor oil is a necessity.   (G.5)

175. All of the following part numbers would appear in an alphanumeric list EXCEPT:
    A. 16937X
    B. 172452
    C. 19900B
    D. B12346   (A.4)

176. Parts Specialist A says that a gasoline engine is an internal combustion engine. Parts Specialist B says that a diesel engine is an internal combustion engine. Who is right?
    A. A only
    B. B only
    C. Both A and B
    D. Neither A nor B   (C.1.1)

177. The camshaft:
    A. rotates.
    B. reciprocates.
    C. vibrates.
    D. is stationary.   (C.1.2)

178. Flywheel runout is measured with the use of a:
    A. torque gauge.
    B. feeler gauge.
    C. micrometer.
    D. dial indicator.   (C.1.3)

179. Blue exhaust is an indication of:
    A. a sound engine.
    B. excessive oil consumption.
    C. rich air/fuel mixture.
    D. coolant leaking into the combustion chamber. (C.1.4)

180. What tool is used to measure the power steering rotors?
    A. Dial caliper
    B. Feeler gauge
    C. Six-inch scale
    D. Micrometer (C.2.3)

181. What is the LEAST likely problem if the thermostat is stuck in the open position?
    A. High coolant level in the recovery tank
    B. Overheating
    C. Poor coolant circulation
    D. Loss of coolant (C.2.4)

182. Parts Specialist A says that a missing air filter will affect engine performance. Parts Specialist B says that a clogged fuel filter will affect engine performance. Who is right?
    A. A only
    B. B only
    C. Both A and B
    D. Neither A nor B (C.3.2)

183. Worn turbocharger seals may be noted by:
    A. black smoke from the exhaust.
    B. white smoke from the exhaust.
    C. colorless smoke from the exhaust.
    D. blue smoke from the exhaust. (C.3.3)

184. Where is the fuel pressure tester connected on most port fuel injection engines?
    A. In series to the fuel inlet line
    B. Between the fuel tank and pressure regulator
    C. Between the pressure regulator and the injector
    D. To the Schrader valve on the fuel rail (C.3.4)

185. The following are all components of the ignition system EXCEPT:
    A. the battery.
    B. the starter.
    C. the ignition coil.
    D. a spark plug. (C.4.1)

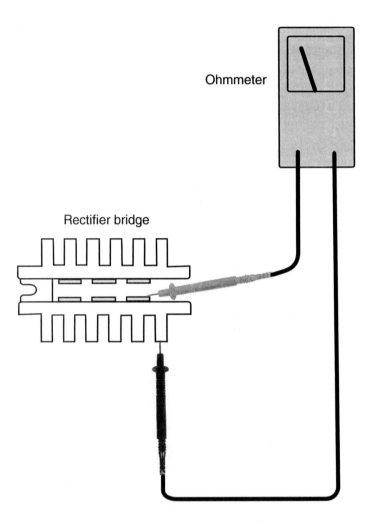

186. Which of the following statements about the test shown in the illustrtion is not true?
    A. The meter hookup is checking a diode.
    B. The meter should read high resistance if the component is good.
    C. The meter should show zero resistance.
    D. The meter should read low resistance if the component is good.   (C.13.4)

187. Which of the following is not found on all vehicles?
    A. Muffler
    B. Exhaust manifold
    C. Exhaust pipe
    D. Resonator   (C.5.2)

188. When checking a PCV valve, air should:
    A. pass freely in either direction.
    B. pass freely in one direction but not the other.
    C. not pass in either direction.
    D. pass intermittently in one direction only.   (C.5.4)

189. The evaporative emission controls (EEC) system:
    A. introduces exhaust gases into the intake air to reduce the formation of oxides.
    B. breaks down oxides of nitrogen into nitrogen and oxygen.
    C. introduces fresh air into the exhaust stream to cause a second burning.
    D. draws vapors from the fuel tank and introduces them into the intake air stream. (C.6.1)

190. Parts Specialist A says that most EGR valves on late-model vehicles are electrically controlled. Parts Specialist B says the EGR valve uses engine vacuum to siphon exhaust gases into the intake manifold. Who is right?
    A. A only
    B. B only
    C. Both A and B
    D. Neither A nor B (C.6.2)

191. All of the following statements about a PCV valve are true EXCEPT:
    A. the PCV plunger is seated in the housing during a backfire.
    B. the PCV plunger is nearly seated at engine idle speed.
    C. a PCV valve plunger stuck in the open position produces increased engine idle speed.
    D. when the PCV valve air filter is clogged, vacuum is reduced in the engine. (C.6.3)

192. Parts Specialist A says that the pressure plate is pulled away from the disc when the clutch is engaged. Parts Specialist B says that the pressure plate and flywheel rotate in opposite directions to each other. Who is right?
    A. A only
    B. B only
    C. Both A and B
    D. Neither A nor B (C.7.2)

193. If the automatic transmission dipstick feels sticky, the indication is:
    A. the fluid has been overheated.
    B. the fluid contains varnish.
    C. damaged transmission components.
    D. a leaky transmission cooler. (C.8.3)

194. Driveshafts may be made of all the following materials EXCEPT:
    A. steel tubing.
    B. aluminum tubing.
    C. fiberglass.
    D. titanium. (C.9.2)

195. Parts Specialist A says constant velocity (CV) joints should be replaced as a complete assembly. Parts Specialist B says that grease included with a repair kit need not be used if the grease in the old joint was sufficient. Who is right?
    A. A only
    B. B only
    C. Both A and B
    D. Neither A nor B (C.9.3)

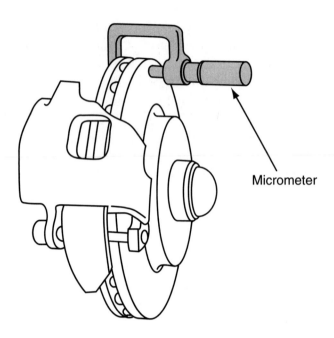

Micrometer

196. Parts Specialist A says the setup is checking the runout of a brake rotor. Parts Specialist B says the setup is measuring the rotor thickness. Who is right?

A. A only

B. B only

C. Both A and B

D. Neither A nor B   (C.10.4)

197. Parts Specialist A says that the front brakes do 75 percent of the work in stopping a vehicle. Parts Specialist B says that rear brakes will provide 50 percent braking capacity in the event that the front brakes fail. Who is right?

A. A only

B. B only

C. Both A and B

D. Neither A nor B   (C.10.2)

198. Parts Specialist A says that English and metric bolts and nuts are interchangeable if the proper type is not available. Parts Specialist B says that a course thread identifies an English bolt while a fine thread identifies a metric bolt. Who is right?

A. A only

B. B only

C. Both A and B

D. Neither A nor B   (C.14.2)

199. Parts Specialist A says that bolt diameter is the measure across the threaded area. Parts Specialist B says that a thread pitch gauge is used to determine thread pitch. Who is right?

A. A only

B. B only

C. Both A and B

D. Neither A nor B   (C.14.3)

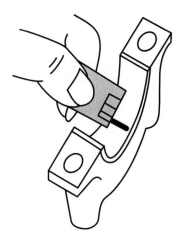

200. The tool shown in the figure is called:
    A. a radius gauge.
    B. plastigage.
    C. a bearing scraper.
    D. a concentricity gauge.   (C.1.4)

Appendices

Answers to the Test Questions
for the Sample Test Section 5

1.	C	31.	B	61.	C	91.	D
2.	B	32.	C	62.	B	92.	B
3.	D	33.	C	63.	A	93.	C
4.	A	34.	A	64.	A	94.	A
5.	B	35.	D	65.	D	95.	A
6.	C	36.	D	66.	C	96.	A
7.	A	37.	C	67.	D	97.	A
8.	C	38.	C	68.	B	98.	C
9.	C	39.	A	69.	C	99.	B
10.	C	40.	D	70.	C	100.	D
11.	C	41.	C	71.	B	101.	B
12.	B	42.	A	72.	D	102.	C
13.	B	43.	D	73.	B	103.	A
14.	D	44.	B	74.	C	104.	D
15.	B	45.	C	75.	A	105.	B
16.	D	46.	D	76.	A	106.	B
17.	D	47.	B	77.	C	107.	C
18.	D	48.	D	78.	A	108.	D
19.	A	49.	B	79.	C	109.	D
20.	C	50.	B	80.	C	110.	D
21.	C	51.	A	81.	B	111.	B
22.	A	52.	C	82.	C	112.	B
23.	C	53.	A	83.	D	113.	A
24.	B	54.	C	84.	C	114.	C
25.	C	55.	C	85.	C	115.	D
26.	D	56.	D	86.	A	116.	A
27.	A	57.	D	87.	A	117.	B
28.	C	58.	D	88.	B	118.	B
29.	C	59.	B	89.	A		
30.	D	60.	A	90.	A		

Explanations to the Answers for the Sample Test Section 5

1. Both parts specialists are correct. A part purchased for $10.00 and sold for $15.00 generates a 50 percent profit. Also, a 10 percent discount on a $25.00 sale is $2.50. **Answer C is correct.**

2. A is wrong, a 5 percent restocking fee would not cost the customer $95.00. A 5 percent restocking fee would cost the customer $5.00; therefore B is right. C is wrong, 100.00 dollars is what the part cost. D is also wrong; the customer paid $100.00 and would not receive more money back than was paid. **The correct answer is B.**

3. Both parts specialists are wrong; 350 cubic inches is about 5.7 liters and 390 cubic inches is about 6.4 liters. **The correct answer is D.**

4. Only choice A is correct. 369482A is before all of the other numbers. 378654C is after 369482A, 369482B is after 369482A, and 388426A is after 369482A. **The correct answer is A.**

5. **Answer B is correct.** The micrometer reading is 0.184 inch.

6. Both parts specialists are right. You must be familiar with the charge policies before making a charge transaction. Also, charge accounts are typically for large sales. **C is the correct answer.**

7. Only Parts Specialist A is right and **the correct answer is A.** The ability to communicate with others is essential for anyone in any business. Parts Specialist B is wrong; it can be very difficult to interact with some people. This is why effort must be made to be patient and understanding, regardless of the customer's behavior.

8. Both specialists are correct. Although the store or parts department may have an employee or cleaning crew that have the responsibility for cleaning the place, a counterperson also has the responsibility for keeping the place tidy and professional looking. **The correct answer is C.**

9. **The correct answer is C.** Both parts specialists are right. An experienced employee is often asked to help train a new employee. It is also better to ask for help than to sell the customer the wrong part.

10. Answer A is the sale price of the motor. This is the amount that will be reimbursed to the customer plus the tax paid on that amount. Answer A is wrong. Choice B is also wrong; this is the store's recorded cost for the motor. Choice C is correct; this is the sale price with tax added. The customer should receive this amount **and the correct answer is C.** Choice D is the cost plus tax. To totally understand the answer, you may want to calculate the tax rate. To do this, divide the tax paid by the total of the parts (9.50 divided by 158.40). You will find the tax to be 6%. Then if you add 6% to the sale price of the motor, you will have a total of $92.54.

11. Answer C is correct. Always evaluate the task before attempting to lift an object. If the part is too heavy, then ask for assistance. Choices A, B, and D are wrong. You should always evaluate the task before asking for assistance. If the part is too heavy, then ask for help and be willing to help others lift a heavy load. Also, before you go ahead and use a hand truck, make sure the part is too heavy for lifting. **The correct answer is C.**

12. The EPA and OSHA have strict regulations on solvents, cutting oils, and caustic cleaning compounds. There are no EPA or OSHA regulations on normal floor soaps. **The correct answer is B.**

13. Parts Specialist A is wrong; large items should not be stored on the top shelf due to the chance that the item could fall and injure someone. Parts Specialist B is right; expensive items should be displayed behind the counter in locked cases to discourage theft. **The correct answer is B.**

14. Answer D is correct. Do-it-yourselfers often need good advice on parts and methods. Do-it-yourselfers usually need more information than professional automotive customers, but they do not always require technical assistance. They usually need good advice on parts and methods but do not always purchase used or rebuilt parts. They also usually need good advice on parts and methods, but seldom do they attempt major engine or transmission repairs. **The correct answer is D.**

15. The tester shown in the figure is a standard cooling system tester. It is used to check for leaks throughout the system, including the radiator cap. The tester is used to create a high pressure in the system or on the radiator cap. If there is a leak, the pressure will not build up. In the case of a radiator cap, pressure should be able to build up to the pressure rating of the cap. If the cap cannot hold a lower pressure, it is bad and should be replaced. The tester does not affect the temperature; therefore, only Parts Specialist B is right. **The correct answer is B.**

16. Both parts specialists are wrong. A parts specialist's opening question should be designed to put the customer at ease and elicit as much data as possible. Greeting with the question, "May I help you?" gives the customer an easy opportunity to reply, "No." **The correct answer is D.**

17. **The correct answer is D** because both parts specialists are wrong. A parts specialist should not be concerned with placing blame because it is not always evident that there is blame to place. Also, a parts specialist should take whatever time necessary to understand the customer's viewpoint and should never raise his or her voice.

18. Both parts specialists are wrong. The first step toward making a customer feel noticed is to make eye contact. Do this when greeting the customer and when talking with him or her. From the time a customer walks through the door until leaving the store, the person should be acknowledged and responded to. **The correct answer is D.**

19. **The correct answer is A.** Parts Specialist A is right; the parts specialist should politely ask the person at the counter to wait while the parts specialist answers the phone. Keep the phone conversation as brief as possible and continuously acknowledge the person at the counter.

20. To measure valve spring free height, the spring is removed from the valve assembly and measured from top to bottom. Answer B is correct for dimension B, not A. This dimension represents installed valve spring height. Answer C is correct. Dimension A shows the measurement for installed valve stem height. Choice D is wrong because there is no true clearance between the retainer and the seat. The spring occupies that distance. **The correct answer is C.**

21. Both parts specialists are right and **the correct answer is C.** Customers are generally influenced by a clean, neat, parts specialist and by the condition of the store as well as the appearance of the employees.

22. Parts Specialist A is right; selling related parts can often boost profits by as much as 30 percent. In fact, "up-selling" is a characteristic of a great counterperson. Parts Specialist B is wrong; a parts specialist should not force a sale by promoting slow-selling merchandise. **The correct answer is A.**

23. If you understand the major components of an A/C system and look carefully at the arrangement of the parts, you should be able to identify the parts. Plus as a hint, the drawings give you an outline of the shape of the component. In the drawing, component 1 is the compressor, 2 is the condenser, 3 is the evaporator, 4 is the accumulator, and 5 is a pressure switch. Both parts specialists are correct and **the correct answer is C.**

24. The component is a VIR. If you look carefully you will note there is a desiccant bag like a typical receiver/drier, but there are also control valves in the unit. There is a TXV and a POA valve. This is why it is called a valves-in-receiver unit. **The correct answer is B.**

25. The part shown is an ignition coil for a distributorless ignition system. Since there are three terminals showing, this coil pack will serve at least three cylinders. Perhaps the terminals for the other cylinders are not noticeable in the drawing as there are few three-cylinder engines. Therefore both parts specialists are correct and **the correct answer is C.**

26. This is a tricky question, or so it seems. You are asked to identify the part that is not part of the fuel system. The fuel pump, fuel tank, and carburetor are part of the fuel system. The fuel tank hangers help support and secure the fuel tank in the vehicle but are not a part of the fuel system. **The correct answer is D.**

27. Parts Specialist A is right; vacuum operated EGR valves have a poppet valve and a diaphragm. They are also part of the emissions control system. They recirculate a small amount of exhaust into the combustion chambers to control the temperatures that are reached. Reducing the temperature reduces the amount of NO_x in the exhaust. **The correct answer is A.**

28. The figure is that of a typical EGR valve. An EGR (exhaust recirculation valve) sends a sample of the engine's exhaust into the combustion chamber to reduce the temperature formed by combustion. Doing this reduces the amount of NO_x that is released in an engine's exhaust. NO_x is one of the controlled exhaust emissions and is known to cause smog. Therefore both parts specialists are correct and **the correct answer is C.**

29. The short in the drawing is after the load. It completes the circuit for the lamp causing it to be on always. The switch serves the function of only completing the circuit to ground; therefore, with the short to ground the switch would be useless. A short after the load would not cause amperage to increase and therefore the fuse would not blow, nor would the wire melt or burn. Since there is not an increase in amperage, the switch's contacts would not arc. **The correct answer is C.**

30. The meter shown is connected across the ground circuit of a starter motor. This hookup would display the voltage drop across the ground. In an ideal world, this reading should be zero. A voltage drop of 0.2 volts would be slightly high and would indicate excessive resistance in the starter ground circuit. Both specialists are wrong and **the correct answer is D.**

31. The part shown is a ball joint. A ball joint allows the control arms to move up and down which keeps the tires on the road. **The correct answer is B.**

32. Both parts specialists are right. A vehicle's suspension system helps to cushion the vehicle from road shocks and helps to keep the vehicle in contact with the road or driving surface. **The correct answer is C.**

33. Both specialists are correct. Camber is a tire wearing angle. Camber changes at the level or height of the suspension changes. Specifications for camber provide good tire contact as the suspension responds to the road surfaces. If the strut is damaged, camber will be wrong. **The correct answer is C.**

34. **The correct answer is A** because Parts Specialist A is right; the starter motor turns a flywheel that is mounted on the rear of the crankshaft. Parts Specialist B is wrong; when the ignition key is turned to the START position, battery voltage is applied directly to the starter motor, not to the alternator.

35. Bolt dimensions are determined by the diameter and pitch of the bolt threads. The bolt length is also used in stating dimensions. Answer D is correct. Choices A, B, and C are wrong. The distance across the flats of a bolt head determines the size wrench to use; likewise, the correct wrench size to use is determined by the distance across the flats of a bolt head. C is wrong because the distance across the points of a hex head have nothing to do with a bolt's thread diameter. **The correct answer is D.**

36. Both specialists are wrong. The standard location for the vehicle identification number is attached to the driver side of the instrument panel, visible through the windshield. **The correct answer is D.**

37. Answer A is wrong. The production date is not found on the trunk lid. Nor is it found on the passenger side rear door, as suggested by choice B. Nor is it found on the passenger side inner fender under the hood, as claimed in choice D. It is, as suggested by C, found on the driver's door or sill plate. **The correct answer is C.**

38. **The correct answer is C** because it is not a true statement. The driveshaft does not have a tag; therefore it would not use a similar tag. Tagging the driveshaft may cause it to run out-of-balance and cause driveability problems. The engine, transmission, or axle may have a tag like the one in the illustration.

39. The procedure is called bench bleeding a master cylinder. This procedure is recommended prior to installation of the unit to a vehicle. Bench bleeding removes the air from the internal parts of the master cylinder. Failure to do this will make bleeding of the brakes, after installation, very difficult. It is not a substitute for regular brake bleeding. Anytime the brake system is opened, the entire system needs to be bled. Fluid is put in the master cylinder's reservoir and the dowel is used to move the pistons in the cylinder. This movement pumps the fluid through the cylinder and pushes the air out. The tubes and dowel must be removed prior to installation. Parts Specialist B is wrong and **the correct answer is A.**

40. Both parts specialists are wrong and **the correct answer is D.** The paint code is located on a separate trim tag and it is not found in the owner's manual. The paint code is not part of the vehicle identification number (VIN). It is located on a trim tag, called the service parts identification label, which is mounted under the hood or on the driver's door of some vehicles.

41. Both parts specialists are right. Each catalog has, in the upper right-hand corner, the form number, the date of issue, and the number and date of the catalog it replaces or supplements. The cover also identifies the product line and manufacturer and often provides an area where the jobber can stamp his or her name for the convenience of dealer customers. **The correct answer is C.**

42. Parts Specialist A is right; footnotes are often very important for getting the correct part for the customer. Footnotes give additional information to determine exceptions or to find alternative parts. Sometimes footnotes are difficult to understand when a number of them are squeezed on a page. The footnotes themselves are important. Footnotes should be referenced systematically in order to fully define those applications that have multiple possibilities. **The correct answer is A.**

43. **The correct answer is D.** Flare, compression, and pipe type fittings are used in automotive service. Answer A is wrong. A tubing type fitting is not used on automobiles. Answer B is also wrong. A steel type fitting is not used on automobiles. Also, choice C is wrong. Neither steel nor tubing type fittings are used in automobiles.

44. Bulletins are lists of updated information that manufacturers send to jobbers between issues of their catalogs. They are the primary tool for conducting catalog maintenance, which involves continually updating and revising the catalog racks and making sure counter personnel are using the most up-to-date information from those catalogs. Bulletins are usually one of the following types: (1) new item availability bulletins which list items now in stock; (2) supersession bulletins, which note part numbers that now supersede previously noted numbers; (3) product information bulletins; which contain specific information about a product, such as a manufacturing defect or a unique method of installation. (4) technical bulletins, which alert counter personnel to any unusual installation or fit problems or unique maintenance tips; and (5) correction bulletins, which refer to catalog errors due to mistyping or inaccurately assigned parts numbers. **The correct answer is B;** there are no delivery bulletins.

45. **The correct answer is C.** In the figure, AC is the abbreviation for air conditioning.

46. An impulse product is a product that the customer buys on the spur of the moment to fill a "want" rather than a "need" for the item. Some examples of impulse items include vehicle appearance-enhancing items such as pin striping kits, chrome accessories, high-flow air filters to improve vehicle performance, and interior items such as extra cup holders or seat covers. **Answer D is correct.** The other answer choices are not impulse items.

47. **The correct answer is B.** Parts Specialist A is wrong. Generally, all stock should be rotated so other items are not continually pushed to the rear of the shelf. Parts Specialist B is right. Displaying the price on a shelf helps the overall appearance of the store. The customer does not have to move the merchandise in order to determine its cost, thus maintaining neater shelf displays.

48. This is an except-type question. All of the answer choices are true statements about display pricing except D and **the correct answer is D.** Displaying an item does not mean that it will always be on sale. Display pricing is often used for introduction of new items and special sale items. The customer is more likely to notice the price and product in a display than on the shelf.

49. **The correct answer is B;** a display arrangement in an open area is most effective. Answer A is wrong because there is no regulation on height. Choice C is also wrong; display arrangements should be placed in the open away from the usual location for the product. And choice D is wrong. Displays can be set up by anyone.

50. **The correct answer is B.** Four liters is almost equal to four quarts. One quart is equal to 0.9464 liters, so one quart is almost equal to one liter.

51. An invoice should be completed for each sale in the store. These invoices are used to track inventory and customer purchases for billing purposes. A stock order is used by the store to order more stock from the suppliers. A back order refers to merchandise ordered from a supplier but not shipped, due to the supplier being out of stock. Parts Specialist A is right; an invoice should be completed for each and every sale. Parts Specialist B is wrong; a back order refers to merchandise ordered from a supplier but not shipped due to the supplier being out of stock. **The correct answer is A.**

52. Both parts specialists are right. Some customers simply lack confidence rather than skill. Providing information can boost confidence. The parts specialist's greatest impact on customers' confidence is his or her own attitude about their abilities. Avoid speaking down to customers, and converse with them according to their actual abilities. Some problems may be beyond the capabilities of some do-it-yourselfers. **The correct answer is C.**

53. Defensive selling implies a method of selling that protects the interests of the store in response to do-it-yourselfers who are not always clear and methodical in their diagnosis of a problem. These customers will sometimes purchase the wrong part, install it, find that the problem has not been solved, and then try to return the part as defective. Parts that are sold and returned in this manner are so done at the expense of the store. For this reason, many stores have a policy that does not allow the return of parts that have been installed. Most stores require that parts be returned in their original containers. Therefore Parts Specialist A is right. Parts Specialist B is wrong; do-it-yourselfers may not have enough knowledge and diagnostic skills to replace the correct part. **The correct answer is A.**

54. When the parts specialist carries many different brands of the same product, the difference between them should be explained. If the store carries, for example, four different kinds of brake pads, the parts specialist should identify the material, size, cost, warranty, and name brand differences. The parts specialist should let the customer decide on which part to buy. The parts specialist can make recommendations, but the final decision should be the responsibility of the customer. **The correct answer is C.**

55. After some experience, a parts specialist will know when a customer is ready to buy and when a customer is there to browse. A parts specialist should not push the sale of any parts that a customer may not need. Part of a parts specialist's job is to be able to identify the different customer types and determine their needs. Both parts specialists are right and **the correct answer is C.**

56. **The correct answer is D** because both parts specialists are wrong. If a parts specialist must put a telephone customer on hold, he or she should identify the name of the store and politely state something such as, "Please hold one moment." The parts specialist should never pick up the phone and press the hold button without saying anything, and he or she should never push the button before finishing a sentence or phrase. If the parts specialist knows that it will take a while before he or she can talk with the caller, the parts specialist should ask that the caller call back rather than be put on hold for a long period of time.

57. A good parts specialist should always promote sales and store services. If a customer asks for some brake pads, the parts specialist should let the customer know that they have a machine shop and can turn his rotors or drums if needed. The parts specialist can also let the customer know that they have brake fluid and other related brake parts in stock to further promote the store sales. Only choice D is correct and **the correct answer is D.**

58. Both Parts Specialists A and B are wrong. **Answer D is correct.** A good parts specialist should always promote sales and store services. If a customer asks for some brake pads, the parts specialist should let the customer know that they have brake fluid and other related brake parts in stock to further promote the store sales. Selling related items is extremely important, not only to build profits, but also to help the customer achieve the safest possible results.

59. A parts specialist should help the customer get the right part each and every time. If a customer comes in for a starter while carrying a pair of jumper cables, the parts specialist might ask, "What was your car's problem that lead you to believe that it needs a starter?" The customer may say, "It will not start without jump-starting the battery." The parts specialist should then explain that the battery could be the problem. This way the parts specialist will not only get the sale for the battery but will not have sold the customer a part that did not fix the vehicle. **Answer B is correct.**

60. **The correct answer is A.** Closing a sale means getting a commitment from the customer. Taking the customer's money follows after the commitment. Looking up the correct part occurs before the customer makes a commitment. Handing the customer their change and receipt occurs after taking their money, which follows closing the sale.

61. Both parts specialists are right and **the correct answer is C.** A vehicle build sheet can indicate the engine size and the color of the vehicle.

62. Parts Specialist A is wrong; the index, not the table of contents, will show the catalog sequence. Parts Specialist B is right and **the correct answer is B.**

63. Parts Specialist A is right. Bulletins are the primary tool for conducting catalog maintenance, which involves continually updating and revising the catalog racks and making sure counter personnel are using the most up-to-date information from those catalogs. **The correct answer is A.**

64. As with any delivery, an order for a shipment must be carefully picked, packaged, and documented. A packing list or slip must be included with each shipment. This list contains a detailed description of the items included in the shipment. A separate packing list can be placed in each part of a multiple-part shipment, but often a single comprehensive list is packaged with or secured to the first part of a multiple-part shipment. Parts Specialist A is right; a packing list contains a detailed description of the items included in the shipment. Parts Specialist B is wrong; a packing list contains a detailed description of the items included in the shipment but no directions for where to deliver the package. **The correct answer is A.**

65. **Answer D is correct.** Physical inventory should be done once a year.

66. **Answer C is correct.** Physical inventory discrepancies are reported by filling out an inventory discrepancy form.

67. Parts Specialist A is wrong; stock rotation prevents older stock from sitting at the back of the shelf and eventually having an aged appearance. Parts Specialist B is also wrong. Stock should be rotated every time that the shelves are stocked. Both specialists are wrong and **the correct answer is D.**

68. **Answer B is correct.** A special order is a part that is ordered because the store does not keep the part in stock.

69. **Answer C is correct.** A core charge is a refundable charge that is added when a customer buys a remanufactured part.

70. Both specialists are right and **the correct answer is C.** Additional forms must usually be completed for warranty-return parts. This allows the manufacturers to determine what is wrong with the part and to credit the store for the return. A warranty-return part should never be placed back in the inventory to be sold to another customer.

71. **Answer B is correct.** A case refers to the number of parts packaged together and sold as a unit.

72. **Answer D is correct.** A restocking fee would be charged to a customer for returning a part for a refund.

73. Parts Specialist A is wrong; lost sales reports are beneficial to the store. Parts Specialist B is right; the main benefit of using lost sales reports is the ability to spot trends in the aftermarket parts business. **The correct answer is B.**

74. Both specialists are right. Flywheel runout causes an uneven clutch plate mating surface resulting in clutch grabbing. The pressure plate must be installed in its original position to maintain proper balance. **The correct answer is C.**

75. **Answer A is correct.** A burned exhaust valve causes a "puff" noise in the exhaust. Answer B is wrong; high fuel pressure causes a rich air/fuel ratio. Answer C is also wrong. A restricted return fuel line causes a rich air/fuel ratio. Choice D is also wrong. A sticking fuel pump check valve may cause hard starting.

76. All of the statements about oil pump clearance measurements are true except choice A and **A is the correct answer.** If the clearance between the rotors is normal, there is no need to measure the inner rotor diameter. Measuring the clearance between the rotors, the thickness of the inner and outer rotors, and the clearance between the outer rotor and the housing are valid oil pump measurements.

77. Both specialists are right. The water pump bearing may become contaminated by coolant leaking past the pump seal. A defective water pump bearing may cause a growling noise at idle speed. **Answer C is correct.**

78. Parts Specialist A is right; an intake manifold vacuum leak may cause a cylinder to misfire at idle speed. Parts Specialist B is wrong; an intake manifold vacuum leak will not cause a cylinder to misfire during acceleration when the manifold vacuum is reduced. **Answer A is correct.**

79. Both specialists are right. Part number 8 is a cable assembly that moves the self-adjusting lever on the star adjuster when the brake shoes expand as the brakes are applied. Part number 17 keeps the shoes aligned and prevents them from dropping down as the shoes are applied to the drum. **The correct answer is C.**

80. In this except-type question, choice C is wrong. The mechanical advance rotates the reluctor ahead of the distributor shaft in the same direction as shaft rotation. The vacuum advance does rotate the pickup plate in the opposite direction to shaft rotation. The vacuum advance does control spark advance in relation to engine load. The mechanical advance does control the spark advance in relation to engine rpm. **Answer C is correct.**

81. When diagnosing a no-start condition, a 12V test light may be connected from the negative primary coil terminal to ground. If the test light flashes while cranking the engine, the primary circuit is being triggered off and on by the module. A problem in the secondary ignition circuit may be preventing the spark plug from firing. If a test spark plug, connected from each spark plug wire to ground, does not fire when cranking the engine, check the coil, distributor cap and rotor, spark plugs, and wires. If the test light does not flash while cranking the engine, a defective module or pickup coil is indicated. Parts Specialist A is wrong; with the test spark plug connected to the coil secondary wire, the cap and rotor are not subjected to secondary voltage and cannot affect this voltage. Parts Specialist B is right in saying since the 12V test light is flashing while cranking, the primary circuit is triggered on and off. If the coil does not fire the test spark plug, the coil must be defective. **The correct answer is B.**

82. Both parts specialists are right and **the correct answer is C.** The positive crankcase ventilation (PCV) system may draw unfiltered air into the engine through a loose oil filler cap or a leaking rocker arm cover gasket. Leaks in gaskets, such as rocker arm cover or crankcase gaskets, will result in oil leaks and the escape of blowby gases to the atmosphere. The positive crankcase ventilation (PCV) system also draws unfiltered air through these leaks into the engine. This may result in early wear of engine components when the vehicle is being operated in dusty conditions.

83. This is an except-type question. All of the answer choices are true statements about manifold heat control valves except for choice D. **Answer D is correct.** A manifold heat control valve stuck in the closed position increases (not reduces) intake manifold temperature. A manifold heat control valve stuck in the closed position causes a loss of engine power. A manifold heat control valve stuck in the open position may cause an acceleration stumble. A manifold, heat control valve improves fuel vaporization in the intake manifold, especially when the engine is cold.

84. **The correct answer is C.** This is not a true statement. The part is an outboard CV joint. It is being removed from an axle or half shaft. The joint normally comes with a pack of lubricant that must be inserted over and in the joint during installation. A boot is installed over the joint to keep the lubricant and joint free of moisture and dirt. Contamination will cause premature joint failure.

85. **The correct answer is C** because both parts specialists are right. In the upstream mode, the powertrain control module (PCM) energizes both the secondary air injection bypass (AIR Bypass) and secondary air injection diverter (AIR Diverter) solenoids, and vacuum is supplied to both the AIR Bypass and AIR Diverter valves. If the voltage is 12V on both terminals of the AIR Diverter valve, there is no voltage drop across the winding because there is no current flow through the winding. This condition may be caused by an open wire from this solenoid to the PCM, or the PCM may not be grounding the wire.

86. **The correct answer is A.** Most linkage adjustments are performed with the gear shift lever in neutral.

87. A five-speed manual transaxle has a growling and rattling noise in third gear only. The cause of the problem must be something that affects only third gear and only affects that gear when the transaxle is operating that gear. **Answer A is correct.** Worn or chipped teeth on the third speed gear could result in a growling noise while driving in third gear. Answers B and C are wrong. Worn dog teeth on the third speed gear may cause hard shifting or jumping out of third gear, but this defect would not cause a growling noise while driving in third gear. Choice D is also wrong. Worn threads in the third speed blocking ring may cause hard shifting, but this wear would not cause a growling noise in third gear.

88. Parts Specialist A is wrong; since the rear planetary gearset is not turning with the engine running and the vehicle stopped, a defective rear planetary gearset would not cause this noise. Parts Specialist B is right; since the oil pump is turning continually with the engine running, this component may be the cause of the whining noise. **Answer B is correct.**

89. Parts Specialist A is right; a misadjusted manual valve shift linkage may cause low fluid pressure that may cause clutch slipping and premature failure. Parts Specialist B is wrong; a misadjusted manual valve shift linkage may cause low fluid pressure that may cause clutch slipping and premature failure. **The correct answer is A.**

90. Answer A is correct. Parts Specialist A is right; a clunking noise while decelerating may be caused by a worn inner drive axle joint. Parts Specialist B is wrong; a worn front wheel bearing usually results in a growling noise while cornering or driving straight ahead. **The correct answer is A.**

91. All of the statements about differential case and ring gear assembly removal and replacement are true except D; the side bearings should be lubricated before installation. The other answer choices are right. Ring gear runout should be measured before removing the ring gear and case assembly. Case side play should be measured before removing the ring gear and case assembly. And side bearing caps should be marked in relation to the case prior to removal. **D is the correct answer.**

92. Parts Specialist A is wrong; damaged brake lines should be replaced, not repaired by inserting a short tube to replace the damaged section. Parts Specialist B is right; a tube-bending tool should be used to make the necessary brake tubing bends without kinking the brake tubing. **The correct answer is B.**

93. This is the normal setup for checking spring squareness and freestanding height. If the spring is distorted, it will not sit squarely on the tool. **The correct answer is C.**

94. **Answer A is correct.** When one side of the bumper is pushed downward with considerable force and then released, the bumper should only complete one free upward bounce if the shock absorber or strut is satisfactory. The other answer choices are wrong.

95. Parts Specialist A is right; harsh riding, excessive steering effort, and rapid steering wheel return may be caused by excessive positive caster. Parts Specialist B is wrong; an incorrect included angle would not cause these symptoms. **Answer A is correct.**

96. The meter shown is connected across the ground circuit of a solenoid. This hookup would display the voltage drop across the ground. To measure voltage at the starter, the meter should be connected so the positive lead is on the starter battery terminal and the negative lead on the negative post of the battery. To measure the voltage drop across the solenoid, connect the positive lead to the battery terminal at the solenoid and the negative lead at the ground of the starter. Battery voltage is measured by connecting the leads of the meter across the battery. **The correct answer is A.**

97. **The correct answer is A** because Parts Specialist A is right. Restricted evaporator refrigerant passages may cause frosting of the evaporator outlet pipe. Parts Specialist B is wrong; this condition may cause lower, not higher, than specified low-side pressures.

98. Both specialists are right. The thermal expansion valve controls the refrigerant flow to the evaporator in response to the temperature of the refrigerant line. The remote bulb senses that temperature and changes the opening of the valve in correspondence to the temperature. This effectively keeps the system operating efficiently. **The correct answer is C.**

99. When testing a diode, connect the ohmmeter leads across the diode and then reverse the leads. A good diode will have one high and one low ohmmeter reading. **The correct answer is B.** Answer A is wrong; two high meter readings indicate a shorted diode. Choice C is wrong. Two low meter readings indicate an open diode. And choice D is wrong; a meter reading of 0 Ohms and 10 Ohms indicates a defective diode.

100. Both parts specialists are wrong. The best location to attract impulse buying is between the customer's chest and eye level. **Answer D is correct.**

101. The answer to this one should be quite obvious. The most common method used to cool an engine is to circulate a liquid coolant through the engine block and cylinder head. Liquid cooling is preferable to air cooling because it is less noisy and is better able to maintain a constant temperature at the cylinders. It also lets the engine operate more efficiently and makes a ready supply of hot coolant available to operate a heater for the in-vehicle compartment. **Answer B is correct.**

102. Both Parts Specialists A and B are right. The fuel system supplies a combustible mixture of gasoline and air to the engine cylinders. In order to do this, it must store the fuel and deliver it to the fuel metering and atomization system, where it is mixed with air to provide the combustible mixture that is delivered in a manner that meets the varying load requirements of the engine. Parts Specialist A is right; fuel must be atomized for proper combustion. Parts Specialist B is also right; air is mixed with the fuel for atomization. **The correct answer is C.**

103. **The correct answer is A** because only Parts Specialist A is right. There is more than one coil in a distributorless ignition system (DIS). Parts Specialist B is wrong; the ignition module synchronizes the coil in relation to the crankshaft position and firing order. The computer, ignition module, and position sensor combine to control spark timing and advance. The computer collects and processes information to determine the ideal amount of spark advance for the particular operating conditions. The ignition module uses crank/cam sensor data to control the timing of the primary circuit in the coils. Remember that there is more than one coil in a distributorless ignition system. The ignition module synchronizes the coil's firing sequence in relation to the crankshaft position and firing order of the engine. Therefore, the ignition module takes the place of the distributor.

104. The question asks you to identify, from the list, the part that is not part of an exhaust system. A muffler, resonator, and catalytic converter are part of a typical exhaust system. A reverberator is not a part of the exhaust system; therefore **the correct answer is D.**

105. **Answer B is correct.** EPA is the abbreviation for Environmental Protection Agency. This agency is responsible for setting automotive emissions standards. EGR is a J1930 term for Exhaust Gas Recirculation. EEC is an acronym for Evaporative Emission Controls. EFE is a J1930 term for Early Fuel Evaporation.

106. **The correct answer is B.** Parts Specialist B is right; the clutch pressure plate is operated by a clutch release bearing. Parts Specialist A is wrong; pressing the clutch pedal disengages the clutch.

107. **The correct answer is C** because both specialists are right. Parts Specialist A is right; the torque converter provides fluid coupling in the transmission. Parts Specialist B is also right; the torque converter allows maximum slippage at engine idle speed.

108. The tool shown is a special tool used to separate a refrigerant line on some systems. The tool depresses the positive lock and the line fitting can be separated. Neither specialist is correct and **the correct answer is D.**

109. **Answer D is correct.** Both parts specialists are wrong. Specialist A is wrong because the rear brakes provide about 20 percent of the braking capacity if the front brakes are in good order. Parts Specialist B is also wrong; the front or rear brakes will provide 50 percent of the braking capacity if either of the systems fail.

110. **Answer D is correct.** Both parts specialists are wrong. Four-wheel steering is currently available for several types of vehicle.

111. Parts Specialist A is wrong; the refrigerant leaving a compressor is a high-pressure vapor, not a liquid. Parts Specialist B is right. The refrigerant entering the compressor is a low-pressure vapor. **The correct answer is B.**

112. **The correct answer is B** because an alternator or generator is on all modern vehicles and certainly would not be considered an accessory. A radio and sound system, power seats and windows, and rear window defogger are not necessary and can be considered accessories, although they may be standard equipment on some vehicles.

113. **Answer A is correct.** ISO is an abbreviation for International Standards Organization.

114. **Answer C is correct.** Both parts specialists are right. A nut should be the same grade as the bolt it is used on and all bolts securing a part should be of the same grade.

115. **Answer D is correct.** Both parts specialists are wrong. The paint topcoat is about 0.004 inch (0.102 mm) thick when applied to metal or plastic.

116. Only a compression fitting uses a ferrule so **the correct answer is A.**

117. Parts Specialist A is wrong; not all hoses used in automotive service withstand high pressure. Parts Specialist B is right; some hoses used in automotive service are made of a reinforced synthetic rubber. **The correct answer is B.**

118. The meter is checking a diode in the rectifier. A good diode will have high resistance when the meter is connected in one direction. If the leads of the meter are reversed, the meter should show a low resistance reading. **The correct answer is B.**

Answers to the Test Questions for the Additional Test Questions Section 6

1.	D	37.	C	73.	A	109.	C
2.	D	38.	C	74.	B	110.	A
3.	D	39.	B	75.	C	111.	B
4.	A	40.	D	76.	C	112.	C
5.	C	41.	A	77.	C	113.	D
6.	C	42.	A	78.	B	114.	C
7.	A	43.	C	79.	C	115.	C
8.	A	44.	A	80.	B	116.	A
9.	A	45.	A	81.	D	117.	B
10.	A	46.	D	82.	A	118.	A
11.	B	47.	D	83.	C	119.	C
12.	A	48.	C	84.	A	120.	D
13.	C	49.	B	85.	C	121.	A
14.	C	50.	B	86.	B	122.	D
15.	B	51.	A	87.	A	123.	D
16.	A	52.	D	88.	C	124.	A
17.	B	53.	B	89.	A	125.	B
18.	A	54.	A	90.	C	126.	D
19.	C	55.	C	91.	C	127.	A
20.	A	56.	C	92.	A	128.	D
21.	A	57.	B	93.	C	129.	C
22.	C	58.	D	94.	B	130.	C
23.	B	59.	B	95.	B	131.	C
24.	B	60.	A	96.	A	132.	C
25.	A	61.	B	97.	C	133.	C
26.	D	62.	A	98.	A	134.	B
27.	D	63.	C	99.	B	135.	A
28.	C	64.	C	100.	A	136.	A
29.	A	65.	D	101.	A	137.	A
30.	B	66.	A	102.	D	138.	C
31.	A	67.	C	103.	C	139.	D
32.	C	68.	C	104.	C	140.	D
33.	A	69.	D	105.	D	141.	A
34.	D	70.	C	106.	D	142.	B
35.	C	71.	D	107.	C	143.	D
36.	B	72.	D	108.	D	144.	C

145.	A	159.	A	173.	D	187.	D
146.	C	160.	A	174.	B	188.	B
147.	D	161.	D	175.	B	189.	D
148.	A	162.	B	176.	C	190.	B
149.	C	163.	B	177.	C	191.	D
150.	D	164.	B	178.	D	192.	D
151.	B	165.	B	179.	B	193.	C
152.	B	166.	D	180.	D	194.	D
153.	B	167.	C	181.	B	195.	A
154.	D	168.	C	182.	C	196.	B
155.	B	169.	A	183.	D	197.	B
156.	D	170.	A	184.	D	198.	D
157.	D	171.	C	185.	B	199.	C
158.	B	172.	D	186.	C	200.	B

Explanations to the Answers for the Additional Test Questions Section 6

1. **The correct answer is D.** Both parts specialists are wrong. To calculate percent discount: Convert the percent into a decimal number by moving the decimal two places to the left. For example, 15 percent (15%) becomes 0.15. Multiply the decimal number by the original price to get the amount of the discount. As in the question, 15% of $60.00 is $9.00 (0.15 × $50.00 = $9.00). To calculate percent profit: subtract the cost from the selling price. For example, if the selling price is $175.00 and the cost is $100.00, the profit is $75.00. ($175.00 – $100.00 = $75.00). Divide the profit by the cost; $75.00 divided by $100.00 is 0.75. Convert this number into a percentage by moving the decimal two places to the right. 0.75 or 0.50 becomes 75 percent. The profit is 75 percent.

2. The restock fee, usually expressed as a percentage, is the fee charged for having to handle a returned part. To calculate the restocking fee, convert the percent into a decimal number by moving the decimal two places to the left. 10% becomes 0.10. Multiply the decimal number by the original price to get the amount of the restocking fee. 0.10 x $250.00 is $25.00. Subtract the restocking fee ($25.00) from the original price ($250.00) to get the amount of money to be returned to the customer. $250.00 – $25.00 = $225.00. **The correct answer is D.**

3. Neither parts specialist is correct. There are 61 cubic inches to a liter. To convert to cubic inches, multiply the liters by 61 to get cubic inches. For example, a 5-liter engine is the same as a 305 cubic inch engine (5 × 61 = 305). The metric unit of temperature measurement is degrees Celsius (°C). To convert from degrees Fahrenheit (°F) to degrees Celsius using a short formula, first subtract 32 then multiply by 0.555. For example 212°F is equal to 99.9°C (212 – 32 = 180, and 180 × 0.555 = 99.9). **The correct answer is D.**

4. **The correct answer is A.** An alphanumeric listing places a series of numbers and letters in order starting from the left digit and working across to the right. If the numbers are the same, the order sequence continues with the letter A, then B, and so forth.

5. Reading a micrometer is done by approaching the measurement in steps. Looking at the figure in this question you can see the measurement is 0.245. Because this is a 0–1 inch micrometer, the reading must be between 0 and 1. Since the 2 is the last complete unit visible on the horizontal line, the first number is 0.200. One ⅟₂₅-inch mark is visible after the 2, so the second measurement is 1 × 0.025 = 0.025. The horizontal line nearly lines up with the 20 on the vertical line, so 0.020 is the third measurement. Adding up the three measurements (0.200 + 0.025 + 0.020) gives a total reading of 0.245. **The correct answer is C.**

6. Parts specialists interact with people. They work with customers, suppliers, manufacturers, and numerous other segments of the aftermarket industry. The ability to communicate is essential, whether it is explaining the features and benefits of a product to a customer, placing an order with a warehouse distributor, or routing the proper paperwork and data through the various departments in the jobber store. Interacting with others is not always easy. When personalities clash, misunderstandings will occur. A parts specialist or manager must be a problem solver, an astute negotiator, and a person who gains satisfaction through serving and helping others. **The correct answer is C.**

7. **The correct answer is A.** It is the responsibility of all employees to keep all areas of the store clean and orderly. All employees suffer from lost sales if the floors, shelves, and displays are a safety hazard or do not appeal to the customer.

8. Shipping, receiving, and stocking materials require physical exertion. Knowing the proper way to lift heavy materials is important. Always lift and work within your ability and seek help from others when you are not sure if you can handle the size or weight of the material or object. Auto parts, even small, compact components, can be surprisingly heavy or unbalanced. Always size up the lifting task before beginning. **The correct answer is A.**

9. The Environmental Protection Agency (EPA), Occupational Safety and Health Administration (OSHA), and other state and local agencies have strict guidelines for handling these materials. OSHA's Hazard Communication Standard (HCS), commonly called the "Right-to-Know Law," applies to all companies that use or store any kind of hazardous chemicals that workers might come in contact with, including solvents, caustic cleaning compounds, abrasives, cutting oils, and other hazardous materials. Compliance with the law requires a system for labeling hazardous chemicals and maintaining Material Safety Data Sheets (MSDS). These sheets must be made available to employees, informing them of the dangers inherent in chemicals found in the workplace. **The correct answer is A.**

10. Specialist A is correct. This setup is used to see how straight the rotor is. Any deflection shown on the dial indicator shows how un-flat the rotor is. Rotor thickness is measured with a micrometer. **The correct answer is A.**

11. Expensive items should be displayed behind the counter in locked cases. Parts specialists should give a lot of thought to what gets displayed near entrances and exits. In the moment that a parts specialist is busy researching parts in a catalog, a thief can easily slip something under his or her jacket and glide through the door. It is best to keep the areas around the entrances clear or to display only large, heavy, or awkward merchandise near the door. Barrels of oil and bags of floor sweep are safe choices for displays near entrances. **The correct answer is B.**

12. Do-it-yourselfers usually need more information than do professional automotive customers. These customers usually need and expect good advice on parts and methods. They view the parts specialist as an expert, and the parts specialist should utilize product knowledge and catalog skills to give the advice needed. Nothing should be sold to a customer that is not actually needed. One way that a parts specialist can assist do-it-yourselfers without taking too much time away from other customers is to have printed how-to information available. Once the parts specialist determines what work the customer will be doing, the parts specialist can provide something to read while assisting other customers. After the customer reads the information, the parts specialist can clarify some important points and answer questions. Each customer's needs can be met without anyone waiting a long time. Since your job is to assist customers, you should take the time that is needed to help them. However, you should do what you can to control that time. **The correct answer is A.**

13. Both parts specialists are right. A parts specialist's opening question should be designed to put the customer at ease and elicit as much data as possible. The question, "May I help you?" is the worst possible opening because it invites the opportunity for the customer to reply, "No." Instead, the parts specialist should ask, "How can I help you?" Another good opening is, "Can I show you anything in particular, or would you like to browse around first?" These questions restrict answers to the positive and give the customer a way out without having to say, "No." The customer's response will generally indicate whether he or she has a defined objective, is just looking, or is seriously considering a purchase. **The correct answer is C.**

14. A parts specialist should not be concerned with placing blame because it is not always evident whether or not there is any blame to place. The customer is not always right, but the parts specialist must be very careful when making corrections. Embarrassed customers will probably spend their money somewhere else. Often the easiest way for the parts specialist to deal with a difficult situation is to put themselves in the customer's shoes. Most people, including parts counter personnel, are angry when something that was purchased causes a problem. Often the anger is directed toward the person who sold it. The parts specialist should take whatever time is necessary to listen to and try to understand the customer's viewpoint. **The correct answer is C.**

15. The tool being used is a drum brake tool designed to match the diameter of the drum to the expanse of the shoes. This procedure is used to set the shoes prior to installing the drum. This sets the shoes so that they need a minimum amount of adjustment. **The correct answer is B.**

16. From the time a customer walks through the door until the time he or she leaves, that person should be acknowledged and responded to favorably. Counter personnel should refrain from personal conversations with one another whenever a customer approaches the counter. Customers usually do not want to feel like a number or just another face in the crowd. The first step toward making a customer feel noticed is to make eye contact with the customer by looking at him or her and saying something such as, "Good morning. I'll be with you in a moment." Eye contact should also be made when the parts specialist is waiting on the customer by looking up from the parts catalog periodically while talking to him or her. **The correct answer is A.**

17. Often the phone will ring while the parts specialist is serving a customer. If no one else is available to answer the phone, the parts specialist should politely ask the customer to wait and then answer the phone. If it appears that the call will take a while, the parts specialist should explain to the caller that he or she is assisting another customer at the moment. After taking pertinent information, the parts specialist should state that he or she will return the call as soon as possible or that the caller should phone back in a few minutes. That way, neither the caller nor the counter customer is kept waiting very long. Since a caller cannot see the parts specialist's facial expressions, gestures, or body language, all communications must be accomplished with words, tones, timing, and inflection. Parts counter personnel must remember to speak clearly and slowly enough to be understood and must request that the customer do so if not clearly understood. **The correct answer is B.**

18. The appearance of the store, as well as the appearance of counter personnel, influences the customer. A clean, neat parts specialist establishes a positive image with customers, and a clean, neat store is equally important. Although a parts specialist is rarely in charge of decor or layout, he or she can maintain the appearance of the store in several ways: keep a rag handy for wiping the counter so that it is free of dirt and grime from used parts; keep the counter free of clutter such as small parts, notes, paper clips, and other such items; keep the displays organized and well stocked; keep the stock and supplies neatly shelved to improve appearance; and assist with basic clean-up by throwing away empty parts wrappers and labels and by removing empty boxes from sight. **The correct answer is A.**

19. Both parts specialists are right and **the correct answer is C.** If the parts specialist suggests related items on the mechanic's first visit, both the customer and the parts specialist can save a lot of time. The customer will be very appreciative of the parts specialist's initial time investment. With this type of service the customer will more than likely return for parts again. Selling related parts when a customer buys a particular item can boost profits by 30 percent or more. Selling related parts also makes sense when the parts specialist remembers that a customer's problem might not be solved solely by replacing a faulty part if the related hardware or chemicals are not also up to peak performance. This is especially important if a related part is on sale.

20. **The correct answer is A.** The procedure shows using plastigage to check bearing clearance. If the results indicate that there is too little or too great a clearance, the bearing may need to be replaced with one of a different thickness. Shaft out-of-roundness would probably not affect the results of this check.

21. **The correct answer is A;** the part shown is a typical in-line fuel filter.

22. The procedure shown is for installing new threaded inserts into bores that had damaged threads. These inserts are available in most thread pitches and sizes. **The correct answer is C.**

23. **The correct answer is B.** The procedure shown is the typical way to adjust valve lash or clearance on an OHV engine. The lash is measured with a feeler gauge and is adjusted with the adjustment screw. The locknut is used to hold that adjustment.

24. To measure valve spring free height, the spring is removed from the valve assembly and measured from top to bottom. Answer B is correct. This dimension does represent installed valve spring height. Dimension A shows the measurement for installed valve stem height. Choice D is wrong because there is no true clearance between the retainer and the seat. The spring occupies that distance. **The correct answer is B.**

25. The tester shown in the figure is a standard cooling system tester. It is used to check for leaks throughout the system, including the radiator cap. The tester is used to create a high pressure in the system or on the radiator cap. If there is a leak, the pressure will not build up. In the case of a radiator cap, pressure should be able to build up to the pressure rating of the cap. If the cap cannot hold a lower pressure, it is bad and should be replaced. The tester does not create a vacuum. Therefore, choice A is wrong, but for this question **the correct answer is A.**

26. If you understand the major components of an A/C system and look carefully at the arrangement of the parts, you should be able to identify the parts. Also, as a hint, the drawings give you an outline of the shape of the component. In the drawing, component 1 is the compressor, 2 is the condenser, 3 is the evaporator, 4 is the accumulator, and 5 is a pressure switch. A radiator and air pump are not part of an A/C system. Both parts specialists are wrong and **the correct answer is D.**

27. **The correct answer is D** even though it is not a true statement. Component 4, which is an accumulator or receiver/drier, prevents liquid refrigerant from reaching the compressor. If liquid reaches the compressor, piston damage may occur.

28. Both parts specialists are correct. In most receiver/drier assemblies, the desiccant bag is replaceable without invading the A/C system. Eventually these bags become saturated and can no longer remove moisture from the system. Because this unit is a VIR, it controls refrigerant flow to the evaporator as well as serves as a receiver/drier. **The correct answer is C.**

29. The part shown is an ignition coil for a distributorless ignition system. Distributorless ignition systems are also referred to as electronic ignition systems, although this leads sometimes to confusion. Many years ago when mechanical breaker points were replaced by electronics, those systems were also referred to as electronic ignition. Those early systems are very different from the ignition systems used today. **The correct answer is A.**

30. An EGR (exhaust recirculation) valve sends a sample of the engine's exhaust into the combustion chamber to reduce the temperature formed by combustion. Doing this reduces the amount of NO_x that is released in an engine's exhaust. NO_x is one of the controlled exhaust emissions and is known to cause smog. If the valve does not fully close, it will allow exhaust fumes to dilute the air/fuel mixture and will cause a rough idle. Normally the valve is open when the engine can run efficiently with the diluted air/fuel mixture. **The correct answer is B** because it is wrong.

31. The short in the drawing is after the load. It completes the circuit for the lamp causing it to be on always. The switch serves the function of only completing the circuit to ground; therefore, with the short to ground the switch would be useless. A short after the load would not cause amperage to increase and therefore the fuse would not blow. **The correct answer is A.**

32. Both parts specialists are right. The thread pitch of a bolt in the English system is determined by the number of threads there are in one inch of threaded bolt length and is expressed in number of threads per inch. The thread pitch in the metric system is determined by the distance in millimeters between two adjacent threads. To check the thread pitch of a bolt or stud, a thread pitch gauge is used. Gauges are available in both English and metric dimensions. **The correct answer is C.**

33. Only Parts Specialist A is right. The standard location for the vehicle identification number (VIN) is attached to the driver side of the instrument panel and is visible through the windshield. **The correct answer is A.**

34. **The correct answer is D** because both parts specialists are wrong. The standard location for the vehicle identification number (VIN) is attached to the driver side (not passenger side) of the instrument panel and is visible through the windshield. The production date is the date that the vehicle was assembled. The standard location for this information is on a tag affixed to the driver's door by the door latch or on the driver's door sill plate. It certainly would not be found on the spare tire cover.

35. Both specialists are correct. Component identification (ID) data are stamped in the casting or on a tag that is similar to the one shown in the figure and attached to the component. All major components on the vehicle will have ID data attached. **The correct answer is C.**

36. There are three major types of truck cabs. The regular cab is one that has one seating surface and two doors. An extended cab, is larger than a regular cab and has one and a half seating surfaces. Extended cabs may have two, three, or four doors. A crew cab has two full seating surfaces and four doors. The long cab is not a true classification of truck cabs and is wrong, but **the correct answer is B.**

37. Both parts specialists are right and **the correct answer is C.** The paint code is not part of the vehicle identification number (VIN). It is located on a separate trim tag, called the service parts identification label, which is mounted under the hood or on the driver's door of some vehicles.

38. **The correct answer is C** because both parts specialists are right. Each catalog has, in the upper right-hand corner, the form number, the date of issue, and the number and date of the catalog it replaces or supplements. The cover also identifies the product line and manufacturer and often provides an area where the jobber can stamp his or her name for the convenience of dealer customers. If catalogs were not the same size, they would be difficult to organize and would make the job of locating parts very difficult.

39. **The correct answer is B.** A footnote is used for referencing additional information. Every parts specialist will have occasion to refer to footnotes in the catalogs at some point, either to reference additional information, determine exceptions, or find alternative parts. Sometimes footnotes are difficult to understand when a number of them are squeezed on a page. The footnotes themselves are important. Footnotes should be referenced systematically in order to fully define those applications that have multiple possibilities.

40. **The correct answer is D.** Correction bulletins are used to make corrections during catalog maintenance. Correction bulletins refer to catalog errors due to mistyping or inaccurately assigned parts numbers.

41. **The correct answer is A.** The abbreviation DLC means Data Link Connector.

42. Parts Specialist A is right. A seasonal item is something that a parts department may not normally stock during certain times of the year. Some seasonal items are stocked throughout the entire year but not in large quantities. An example of one of these items is windshield washer fluid, which is stocked more heavily in the winter season than the summer season but is still sold throughout the year. Another such example is car batteries, which are stocked all year even though battery failures occur more often during the colder parts of the year. **The correct answer is A.**

43. In order to keep shelves and items stocked and looking appealing to a customer, the shelves must be well maintained. The shelves, as well as the items on the shelves, should not be dirty or dusty. Dusty items will not sell well because nobody wants to buy anything that they may feel is old and has been sitting around forever. The items on the shelves should be replenished on a regular basis to encourage repeat customers to count on you to have the items they need at all times. If shelves are messy and items are not easily found, this will discourage a sale or a customer from revisiting your store or department again. A shelf should be labeled to some extent to let people know which items go where. Displaying a price on a shelf where the item goes will help keep shelves in order because a customer will not have to take the item off the shelf to see how the price may differ from a similar item next to it. A customer may not always return the item to the space where it was removed, causing an unorganized appearance. **The correct answer is C.**

44. **The correct answer is A.** A normal practice of display pricing is to display an item that may be specially priced for that week in an eye-catching display in an open part of the store or department, along with a noticeable sign that displays the special pricing. The product that is specially priced may still remain in its assigned shelf position, but this additional display should let customers know this item is on special. A customer is more likely to notice a display with a posted price rather than notice an item on the shelf in its normal position. This kind of display and pricing is also used for the introduction of new items. A vendor may also choose to put up a display if overstocked in a certain item but choose not to lower the price. This will help promote the item and possibly lead to more sales. Display pricing can also be used to bring customers to the department or store by the distribution of fliers with an explanation and the price, by advertising on a billboard with the price, or signs in a window with the price.

45. The best place for a display is in an open area near the front of the store or department. This ensures that the display will catch the customer's eye as soon as he or she enters. The display should be large and noticeable, and it must look appealing to the customer. This is usually accomplished by building the display out of the product being sold. Often the manufacturer of a product will send a display kit that can be assembled in the store and which may include banners, signs, and shelves. Displays are also often built at the end of an isle where the product is located so the customer will notice the product without entering the isle. Displays can be set up with the intent of the customer taking the actual item from the display. The display should be set up with care to keep it from falling apart as items are removed. A display also needs to be maintained for a neat appearance and to keep it appealing to the customer. **The correct answer is A.**

46. A customer orders six liters of transmission fluid. You should give the customer 6 quarts of ATF. One quart is equal to 0.9464 liters, so one quart is almost equal to one liter. Therefore, 6 liters is approximately equal to 6 quarts. In actual practice, this amount will leave the transmission about ¼ quart low on fluid but within the safe operating range. **The correct answer is D.**

47. Looking at the invoice in the figure, you can see that **D is the correct answer.** Answer A is cost of dressing and choice B is the customer's cost or sales price for the dressing. Answer C is the cost of the belt but not what the customer paid.

48. **The correct answer is C** because both parts specialists are right. Charge account sales encourage large purchases or purchases from customers who do not want to carry cash or do not have the cash. Counter personnel must be familiar with the charge policy in their workplace to ensure accuracy and customer satisfaction.

49. Parts Specialist A is wrong. A parts specialist should seek whatever help is necessary in order to sell the correct parts and satisfy the customer. Specialist B is right; experienced parts specialists are often asked to help train new employees. **The correct answer is B.**

50. **The correct answer is B.** Parts Specialist A is wrong. An 8 percent discount would equal $1.60, not $5.00. Parts Specialist B is correct. A part bought for $32.00 and sold for $48.00 generates a 50 percent (50%) profit. To calculate percent discount: Convert the percent into a decimal number by moving the decimal two places to the left. Then, multiply the decimal number by the original price to get the amount of the discount. To calculate percent profit: Subtract the cost from the selling price. Then, divide the profit by the cost and convert this number into a percentage by moving the decimal two places to the right.

51. **The correct answer is A.** If a customer returns a $50.00 part and is charged a 15 percent (15%) restocking fee, $42.50 is returned to the customer. To calculate the restocking fee, convert the percent into a decimal number by moving the decimal two places to the left. Then, multiply the decimal number by the original price to get the amount of the restocking fee. Now, subtract the restocking fee from the original price to get the amount of money to be returned to the customer.

52. **The correct answer is D.** 10001Z would appear first in an alphanumeric listing. An alphanumeric listing places a series of numbers and letters in order starting from the left digit and working across to the right. If the numbers are the same, the order sequence continues with the letter A, then B, and so forth.

53. The scale on the micrometer shown in the figure shows that the micrometer is an inch-type. Therefore Parts Specialist A is wrong in saying the reading in the figure is 0.560 mm. The numbers were right but the measurement wrong. Parts Specialist B is correct in saying the reading is 0.560 inches. **The correct answer is B.**

54. **The correct answer is A.** Parts Specialist A says that parts specialists have to work with customers, suppliers, manufacturers, and other people in the aftermarket industry. Parts Specialist B is wrong. Parts specialists interact with people. They work with customers, suppliers, manufacturers, and numerous other segments of the aftermarket industry. The ability to communicate is essential, whether it is explaining the features and benefits of a product to a customer, placing an order with a warehouse distributor, or routing the proper paperwork and data through the various departments in the jobber store.

55. It is the responsibility of all employees to keep all areas of the store clean and orderly. All employees suffer from lost sales if the floors, shelves, and displays are a safety hazard or do not appeal to the customer. A parts specialist is not responsible for keeping the shop parts clean or tidy. **The correct answer is C.**

56. **The correct answer is C** because both parts specialists are right. Small parts usually do not weigh much; therefore, there is no need to seek help when handling them. Knowing the proper way to lift and work within one's ability is important for handling parts. Shipping, receiving, and stocking materials require physical exertion. Knowing the proper way to lift heavy materials is important. Always lift and work within your ability and seek help from others when you are not sure if you can handle the size or weight of the material or object. Auto parts, even small, compact components, can be surprisingly heavy or unbalanced. Always size up the lifting task before beginning.

57. The Occupational Safety and Health Administration (OSHA) is an organization that sets the guidelines for companies that use hazardous chemicals. **The correct answer is B.**

58. **The correct answer is D.** Expensive items should be displayed behind the counter in locked cases. Parts specialists should give a lot of thought to what gets displayed near entrances and exits. In the moment that a parts specialist is busy researching parts in a catalog, a thief can easily slip something under his or her jacket and glide through the door. It is best to keep the areas around the entrances clear, or to display only large, heavy, or awkward merchandise near the door. Barrels of oil and bags of floor sweep are safe choices for displays near entrances.

59. Parts Specialist A is wrong; a parts specialist should not sell the customers what they want if it will not fix the problem. Parts Specialist B is correct in saying only the parts that are needed for a particular repair should be sold to the customers. **The correct answer is B.** Do-it-yourselfers usually need more information than do professional automotive customers. These customers usually need and expect good advice on parts and methods. They view the parts specialist as an expert, and he or she should utilize product knowledge and catalog skills to give the advice needed. Nothing should be sold to a customer that is not actually needed.

60. The meter shown is connected across the ground circuit of a starter motor. This hookup would display the voltage drop across the ground. To measure voltage at the starter, the meter should be connected so the positive lead is on the starter battery terminal and the negative lead on the negative post of the battery. To measure the voltage drop across the starter, connect the positive lead to the battery terminal at the starter and the negative lead at the ground of the starter. Battery voltage is measured by connecting the leads of the meter across the battery. **The correct answer is A.**

61. Parts Specialist B is right. "May I help you?" is one of the worst greetings a parts specialist could make to a customer. A parts specialist's opening question should be designed to put the customer at ease and elicit as much data as possible. The question, "May I help you?" is the worst possible opening because it invites the opportunity for the customer to reply, "No." Instead, the parts specialist should ask, "How can I help you?" Another good opening is, "Can I show you anything in particular or would you like to browse around first?" These questions restrict answers to the positive and give the customer a way out without having to say, "No." The customer's response will generally indicate whether he or she has a defined objective, is just looking, or is seriously considering a purchase. Although the question proposed by Specialist A is not wrong, it is wrong to say it is the best. **The correct answer is B.**

62. Parts Specialist A is correct. An embarrassed customer will probably spend their money elsewhere. The customer is not always right, but the parts specialist must be very careful when making corrections. A parts specialist must not be concerned with placing blame because it is not always evident whether or not there is any blame to place. Specialist B is wrong. **The correct answer is A.**

63. **The correct answer is C** because both specialists are right. From the time a customer walks through the door until the time he or she leaves, that person should be acknowledged and responded to favorably. Counter personnel should refrain from personal conversations with one another whenever a customer approaches the counter. Customers usually do not want to feel like a number or just another face in the crowd. The first step toward making a customer feel noticed is to make eye contact with the customer by looking at him or her and saying something such as, "Good morning. I'll be with you in a moment." Eye contact should also be made when the parts specialist is waiting on the customer by looking up from the parts catalog periodically while talking to him or her.

64. Both specialists are right. A telephone customer cannot see the parts specialist's body language so all communication must be accomplished with words, tones, timing, and inflection. Parts specialists must remember to speak clearly and slowly enough to be understood and must request that the customer do so if he or she cannot be understood. **The correct answer is C.**

65. Both specialists are wrong. The part shown is a ball joint. A ball joint allows the control arms to move up and down which keeps the tires on the road. **The correct answer is D.**

66. Although a parts specialist is rarely in charge of decor or layout, he or she can maintain the appearance of the store in several ways: keeping a rag handy for wiping the counter so that it is free of dirt and grime from used parts; keeping the counter free of clutter such as small parts, notes, paper clips, and other such items; keeping the displays organized and well stocked; keeping the stock and supplies neatly shelved to improve appearance; and assisting with basic clean-up by throwing away empty parts wrappers and labels and by removing empty boxes from sight. Parts Specialist A is right and **the correct answer is A.**

67. Parts Specialist A is right. A fuel filter is used to remove dirt or other harmful particles that might be in the fuel. Parts Specialist B is also right. Vehicles are either equipped with a carburetor or fuel injection. **The correct answer is C.**

68. **The correct answer is C** because both parts specialists are right. Depending on the ignition system, the coils may be serviced separately or as a complete unit. The computer, ignition module, and position sensor combine to control spark timing and advance.

69. Neither parts specialist is right. The cooling system is the system that is used to heat the vehicle. The exhaust manifold is made of metal designed to hold up to the extreme temperatures and to expand and contract with the changes in temperature. **The correct answer is D.** The exhaust system is designed to channel toxic exhaust fumes away from the passenger compartment; to quiet the sound of the exhaust pulses; and to burn, or catalyze, emissions in the exhaust. A typical exhaust system has the following components: exhaust manifold and gasket; exhaust pipe, seal, and connector pipe; intermediate pipe(s); catalytic converter; resonator and/or muffler; tailpipe; and hardware items including heat shields, clamps, gaskets, and hangers.

70. **The correct answer is C** because it is not true. Toe is adjusted at the tie rods, not camber. Camber is adjusted by changing shims, rotating eccentrics, or by repositioning the strut mount. The other statements are true. Camber changes as a vehicle moves down the road, incorrect camber will cause excessive tire wear, and it is affected by a variety of damaged suspension parts.

71. Neither specialist is right so **the correct answer is D.** The exhaust gas recirculation (EGR) valve opens to allow engine vacuum to siphon exhaust into the intake manifold. The EGR valve contains a poppet valve that lifts off its seat when a vacuum is applied to the diaphragm. Intake vacuum then siphons exhaust into the engine. Like a positive crankcase ventilation (PCV) valve, the EGR valve is a kind of calibrated vacuum leak. The EGR valve uses ported vacuum as its primary vacuum source. Ported vacuum is used because EGR is not needed at idle.

72. The procedure is called bench bleeding a master cylinder. This procedure is recommended prior to installation of the unit to a vehicle. Bench bleeding removes the air from the internal parts of the master cylinder. Failure to do this will make bleeding of the brakes, after installation, very difficult. Fluid is put in the master cylinder's reservoir and the dowel is used to move the pistons in the cylinder. This movement pumps the fluid through the cylinder and pushes the air out. The tubes and dowel must be removed prior to installation. **The correct answer is D.**

73. Parts Specialist A is right. Front-wheel-drive vehicles have a driveshaft for each front wheel. Parts Specialist B is wrong. The inboard constant velocity (CV) joint is typically a plunger type. **The correct answer is A.**

74. Specialist A is wrong. If this cable is broken, not only will the brake not self-adjust, but there will also be an incorrect clearance between the shoes and the drum. This would result in inefficient braking. Specialist B is right. The spring on the strut bar not only keeps the bar in place, it also keeps the shoes expanded to a certain point. **The correct answer is B.**

75. Specialist A is right. The part is an outboard CV joint. It is being removed from an axle or half shaft. The retaining ring holds the joint to the shaft. Replacement of this type of retainer is always recommended when they are removed. When they are expanded for removal, they lose some of their spring tension. To prevent failure, they should be replaced. Both specialists are right and **the correct answer is C.**

76. Both specialists are wrong. In both cases the spring should be replaced. Heating the spring will cause it to lose tension rendering it near useless. Shims are placed under the spring to set the tension at a specified installed height. If the spring is shorter than it should be, it has collapsed and needs to be replaced. **The correct answer is D.**

77. **The correct answer is C** because both parts specialists are right. A vehicle's electrical system is protected by fuses, circuit breakers, and fusible links. These protection devices will blow or open if there is a shorted circuit.

78. In the English system, the tensile strength of a bolt is identified by the number of radial lines (grade marks) on the bolt head. More lines mean higher tensile strength. In the metric system, tensile strength of a bolt or stud can be identified by a property class number on the bolt head. The higher the number, the greater the tensile strength. A bolt with three radial lines embossed on the head would be a grade 3 bolt. **The correct answer is B.**

79. Both parts specialists are right. Component identification (ID) data are stamped in the casting or on a tag that is attached to the component. All major components on the vehicle will have ID data attached. **The correct answer is C.**

80. The pickup truck shown in the figure for this question has a regular cab, one that has one seating surface and two doors. Therefore Specialist A is wrong. Specialist B is correct in saying that an extended cab vehicle could have two, three, or four doors. **The correct answer is B.**

81. **The correct answer is D.** Correction forms are not included in most catalogs. Most catalogs contain form number(s), date of issue, and the name of product lines and manufacturers.

82. **The correct answer is A.** A footnote in a catalog contains information or refers to additional information, determines exceptions, or finds alternative parts.

83. Both parts specialists are right. A supersession bulletin may include part numbers that supersede previously issued part numbers and technical bulletins may alert counter personnel to any unusual installation or application problems. **The correct answer is C.**

84. Parts Specialist A is right in saying that to verify an abbreviation one should refer to the abbreviation list. Parts Specialist B is wrong. Manufacturers usually include additional aids for using their publications such as an abbreviation list, which can make for efficient use of the catalog. For example, abbreviations can often have more than one meaning. FWD can mean front-wheel drive or four-wheel drive. OD can mean overdrive or outside diameter. The definitions of these, and other abbreviations, can only be determined by checking the abbreviation list provided. **The correct answer is A.**

85. **The correct answer is C.** A seasonal item is something that a parts department may not normally stock during certain times of the year. Some seasonal items are stocked throughout the entire year but not in large quantities. An example of one of these items is windshield washer fluid, which is stocked more heavily in the winter season than the summer season but is still sold throughout the year. Another such example is car batteries, which are stocked all year even though battery failures occur more often during the colder parts of the year. Some other items used year-round but stocked more heavily in the winter include antifreeze, winter windshield wipers, and gasoline additives to prevent gas line freeze. Some items more heavily stocked in the summer include air conditioning refrigerant, windshield sunshades, and wax or polish. An impulse item is a product that the customer buys on the spur of the moment to fill a "want" rather than a "need" for the item. Some examples of impulse items include vehicle appearance-enhancing items such as pin striping kits, chrome accessories, high-flow air filters to improve vehicle performance, and interior items like extra cup holders or seat covers.

86. **The correct answer is B.** Returned items should not be put back on the shelf if they are opened as this will give them a used appearance and make the customer uneasy about purchasing them. Items that are returned unopened can be returned to the shelf if they appear to be in good order. In order to keep shelves and items stocked and looking appealing to a customer the shelves must be well maintained. The shelves, as well as the items on the shelves, should not be dirty or dusty. Dusty items will not sell well because nobody wants to buy anything that they may feel is old and

has been sitting around forever. The items on the shelves should be replenished on a regular basis to encourage repeat customers to count on you to have the items they need at all times. If shelves are messy and items are not easily found, or shelves are always empty from not restocking, often enough this will discourage a sale or a customer from revisiting your store or department again. The next item on a shelf should be pulled forward when the one in front of it is sold. This will give your shelf the appearance of being full and items will be easier to find for the next customer. When items are restocked the new items should be put behind the old ones. This will help keep dust from building up on items. A shelf should be labeled to some extent to let people know which items go where. Displaying a price on a shelf where the item goes will help keep shelves in order because a customer will not have to take the item off the shelf to see how the price may differ from a similar item next to it. A customer may not always return the item to the space where it was removed, causing an unorganized appearance.

87. **The correct answer is A.** A normal practice of display pricing is to display an item that may be specially priced for that week in an eye-catching display in an open part of the store or department, along with a noticeable sign that displays the special pricing. The product that is specially priced may still remain in its assigned shelf position, but this additional display should let customers know this item is on special. A customer is more likely to notice a display with a posted price rather than notice an item on the shelf in its normal position. This kind of display and pricing is also used for the introduction of new items. A vendor may also choose to put up a display if overstocked in a certain item but choose not to lower the price. This will help promote the item and possibly lead to more sales. Display pricing can also be used to bring customers to the department or store by the distribution of fliers with an explanation and the price, by advertising on a billboard with the price, or signs in a window with the price.

88. The LEAST important part of a display is the product on the display. Therefore **the correct answer is C.** The best place for a display is in an open area near the front of the store or department. This ensures that the display will catch the customer's eye as soon as he or she enters. The display should be large and noticeable and it must look appealing to the customer. This is usually accomplished by building the display out of the product being sold. Often the manufacturer of a product will send a display kit that can be assembled in the store and which may include banners, signs, and shelves. Displays are also often built at the end of an isle where the product is located so the customer will notice the product without entering the isle. Displays can be set up with the intent of the customer taking the actual item from the display. The display should be set up with care to keep it from falling apart as items are removed. A display also needs to be maintained for a neat appearance and to keep it appealing to the customer.

89. Pipe fittings rely on tapered threads to prevent leaks and **the correct answer is A.** Pipe fittings are made of brass or iron and can be used with air, water, oil, fuels, and various gases. They may be connected by brass, copper, or iron pipe. Flare fitting applications include refrigeration equipment, air compressors, oil burners, and most any type of machinery. Flare fittings may be used with thin-wall tubing where application requires joints to withstand high pressures (up to 2,800 psi [19,306 kPa]) using 0.030-inch (0.762-mm) wall copper tubing. They may also be used with flareable copper, brass, aluminum, and welded steel hydraulic tubing. Tubing flared at 37 or 45 degrees (as applicable) forms a sound joint resisting mechanical pullout. It seals and remains leak-free even when disconnected and reconnected. Compression fittings eliminate the need to flare, solder, or otherwise prepare tubing before assembly. They are for use where excessive vibration or tube movement is not a problem. There are some applications, such as brake service, where compression fittings are not used. All sealing surfaces are precision machined to ensure reliable and leak proof closures. It should be noted that many variations of fitting combination types are available. These include, but are not limited to, flare to pipe, flare to compression, and compression to pipe. These include male-to-female as well as female-to-male provisions.

90. Both parts specialists are correct and **the correct answer is C.** The standard location for the vehicle identification number (VIN) is attached to the driver side of the instrument panel and is visible through the windshield.

91. **The correct answer is C.** An impulse item is a product that the customer buys on the spur of the moment to fill a "want" rather than a "need" for the item. Some examples of impulse items include vehicle appearance-enhancing items such as pin striping kits, chrome accessories, high-flow air filters to improve vehicle performance, and interior items such as extra cup holders or seat covers.

92. Parts Specialist A is correct in saying that rotating stock on the shelves keeps items from expiring or becoming old-looking or dusty. Parts Specialist B is wrong. Some things do become expired, and the inventory should not look like some parts may be old and outdated. **The correct answer is A.**

93. A customer orders three liters of oil; you should give 3 quarts of oil. **C is the correct answer.** One quart is equal to 0.9464 liters, so one quart is almost equal to one liter. Therefore, 3 liters is approximately equal to 3 quarts.

94. **The correct answer is B.** Parts Specialist A is wrong; 327 cubic inches is equal to about 5.2 liters. Parts Specialist B says that 488 cubic inches is equal to 8 liters. There are 61 cubic inches to a liter.

95. Parts Specialist A is wrong. The markup is 14 cents. Sixty-one cents is the difference between the cost of one nut and the sale price of two. Parts Specialist B is right; the markup on the camshaft is 40 percent. To determine the markup, divide the actual amount of markup (in dollars) by the cost. In this case, the actual markup is $124.85 – 89.18, which equals $35.67. Now divide that amount by $89.18. The answer is 40 percent. **The correct answer is B.**

96. Charge account sales encourage large purchases or purchases from customers who do not want to carry cash or do not have the cash. Counter personnel must be familiar with the charge policy in their workplace to ensure accuracy and customer satisfaction. Parts Specialist A is right. However, Parts Specialist B is wrong. Whether a customer has credit or a charge account does not determine if tax needs to be paid. **The correct answer is A.**

97. The parts specialist should seek whatever help is necessary in order to sell the correct parts and satisfy the customer. Customer satisfaction is built by supplying the right information and the correct parts. Both parts specialists are right and **the correct answer is C.**

98. **The correct answer is A.** An invoice should be completed for each sale in the store. These invoices are used to track inventory and customer purchases for billing purposes. A purchase order is used to allow a company to purchase parts. It will describe the parts and quantity of parts to be purchased along with billing information. A stock order is used by the store to order more stock from the suppliers. A back order refers to merchandise ordered from a supplier but not shipped, due to the supplier being out of stock. An emergency order is an order placed with the supplier on a routine basis, usually weekly or biweekly.

99. A purchase order is used to allow a company to purchase parts. It will describe the parts and quantity of parts to be purchased along with billing information. A stock order is used by the store to order more stock from the suppliers. A back order refers to merchandise ordered from a supplier but not shipped, due to the supplier being out of stock. An emergency order is an order placed with the supplier on a routine basis, usually weekly or biweekly. Of the answer choices, only B is not a type of order. It reflects a status of an order. **The correct answer is B.**

100. **The correct answer is A.** Some customers simply lack confidence rather than skill. Providing information can boost confidence. The parts specialist's greatest impact on customers' confidence is his or her own attitude about their abilities. Avoid speaking down to customers, and converse with them according to their actual abilities. Some problems may be beyond the capabilities of some do-it-yourselfers.

101. **The correct answer is A.** Some customers simply lack confidence rather than skill. Providing information can boost confidence. The parts specialist's greatest impact on customers' confidence is his or her own attitude about their abilities. Avoid speaking down to customers, and converse with them according to their actual abilities. Some problems may be beyond the capabilities of some do-it-yourselfers.

102. Defensive selling implies a method of selling that protects the interests of the store in response to do-it-yourselfers who are not always clear and methodical in their diagnosis of a problem. These customers will sometimes purchase the wrong part, install it, find that the problem has not been solved, and then try to return the part as defective. Parts that are sold and returned in this manner are so done at the expense of the store. For this reason, many stores have a policy that does not allow the return of parts that have been installed. **The correct answer is D.**

103. **The correct answer is C.** This question refers to defensive selling. Defensive selling implies a method of selling that protects the interests of the store in response to do-it-yourselfers who are not always clear and methodical in their diagnosis of a problem. These customers will sometimes purchase the wrong part, install it, find that the problem has not been solved, and then try to return the part as defective. Parts that are sold and returned in this manner are so done at the expense of the store. For this reason, many stores have a policy that does not allow the return of parts that have been installed.

104. **The correct answer is C.** Selling related parts when a customer buys a particular item could boost profits by 30 percent or more. Selling related parts also makes sense when the parts specialist remembers that solely replacing a faulty part, if the related hardware or chemicals are not also up to peak performance, might not solve a customer's problem. For example, if selling a wheel cylinder, suggest brake fluid.

105. Neither parts specialist is correct. Selling related parts makes sense when the parts specialist remembers that solely replacing a faulty part, if the related hardware or chemicals are not also up to peak performance, might not solve a customer's problem. For example, if selling a wheel cylinder, suggest brake fluid. **The correct answer is D.**

106. Both parts specialists are wrong. When the parts specialist carries many different brands of the same product, the difference between them should be explained. If the store carries, for example, four different kinds of brake pads, the parts specialist should identify the material, size, cost, warranty, and name brand differences. The parts specialist should let the customer decide on which part to buy. The parts specialist can make recommendations but the final decision should be the responsibility of the customer. **The correct answer is D.**

107. **The correct answer is C** because both parts specialists are right. When the parts specialist carries many different brands of the same product, the difference between them should be explained. If the store carries, for example, four different kinds of brake pads, the parts specialist should identify the material, size, cost, warranty, and name brand differences. The parts specialist should let the customer decide on which part to buy. The parts specialist can make recommendations, but the final decision should be the responsibility of the customer.

108. **The correct answer is D.** After some experience, a parts specialist will know when a customer is ready to buy and when a customer is there to browse. A parts specialist should not push the sale of any parts that a customer may not need. Part of a parts specialist's job is to be able to identify the different customer types and determine their needs. Both parts specialists are wrong.

109. Both parts specialists are right and **the correct answer is C.** After some experience, a parts specialist will know when a customer is ready to buy and when a customer is there to browse. A parts specialist should not push the sale of any parts that a customer may not need. Part of a parts specialist's job is to be able to identify the different customer types and determine their needs. Some customers do come into the store just to look around or to compare prices.

110. **The correct answer is A.** If a parts specialist must put a telephone customer on hold, he or she should identify the name of the store and politely state something such as, "Please hold one moment." The parts specialist should never pick up the phone and press the hold button without saying anything, and he or she should never push the button before finishing a sentence or phrase. If the parts specialist knows that it will take a while before he or she can talk with the caller, the parts specialist should ask that the caller call back rather than be put on hold for a long period of time.

111. **The correct answer is B.** If a parts specialist must put a telephone customer on hold, he or she should identify the name of the store and politely state something such as, "Please hold one moment." The parts specialist should never pick up the phone and press the hold button without saying anything, and he or she should never push the button before finishing a sentence or phrase. If the parts specialist knows that it will to take a while before he or she can talk with the caller, the parts specialist should ask that the caller call back rather than be put on hold for a long period of time.

112. Both parts specialists are right and **the correct answer is C.** A good parts specialist should always promote sales and store services. If a customer asks for some brake pads, the parts specialist should let the customer know that they have a machine shop and can turn his rotors or drums if needed. The parts specialist can also let the customer know that they have brake fluid and other related brake parts in stock to further promote the store sales.

113. Neither specialist is right. A good parts specialist will always promote sales and store services. If a customer asks for some brake pads, the parts specialist should let the customer know that they have a machine shop and can turn his rotors or drums if needed. They probably do not know there is a machine shop or they are unaware that this service is available to them. The parts specialist can also let the customer know that they have brake fluid and other related brake parts in stock to further promote the store sales. **The correct answer is D.**

114. **The correct answer is C** and both parts specialists are right. Selling related items is extremely important, not only to build profits, but also to help the customer achieve the safest possible results. Displays should be set up to advertise the complete service job. Brake pads, shoes, and hardware once came in fairly plain, dull boxes, but today, most manufacturers have upgraded their packaging graphics to the point where these products can be used to build attractive, effective displays in the front of the store.

115. Selling related items is extremely important, not only to build profits, but also to help the customer achieve the safest possible results. Displays should be set up to advertise the complete service job. Brake pads, shoes, and hardware once came in fairly plain, dull boxes, but today, most manufacturers have upgraded their packaging graphics to the point where these products can be used to build attractive, effective displays in the front of the store. Both specialists are right and **the correct answer is C.**

116. Parts Specialist A is right. The goal is to give the customer the right part every time. Parts Specialist B is wrong. Always make sure you are giving the customer the right part and information, to the best of your ability. **The correct answer is A.**

117. Parts Specialist A is wrong. A parts specialist should help the customer get the right part each and every time. If a customer comes in for a starter while carrying a pair of jumper cables, the parts specialist might ask, "What was your car's problem that lead you to believe that it needs a starter?" The customer may say, "It will not start without jump-starting the battery." The parts specialist should then explain that the battery could be the problem. This way the parts specialist will not only get the sale for the battery but will not have sold the customer a part that did not fix the vehicle. **The correct answer is B.**

118. **The correct answer is A** and Parts Specialist A is right. A good sale closer always lets the customer know that they expect the sale. Closing simply means getting a commitment from the customer. Successful salespersons usually share certain characteristics. First, effective closers expect the sale. The very best closers are certain that they will bring each interview to a successful close. Their confidence might be justified, or it might be the product of an inflated ego, but, whatever the source, that expectation of success often results in sales. Second, good closers always let the customer know that they expect the sale. Finally, many parts specialists do a fantastic job of qualifying prospects, matching benefits to needs, laying a foundation of agreement, handling any objections, and then stopping and waiting for customers to start shelling out their money.

119. Closing simply means getting a commitment from the customer. Successful salespersons usually share certain characteristics. First, effective closers expect the sale. The very best closers are certain that they will bring each interview to a successful close. Their confidence might be justified, or it might be the product of an inflated ego, but, whatever the source, that expectation of success often results in sales. Second, good closers always let the customer know that they expect the sale. Finally, many parts specialists do a fantastic job of qualifying prospects, matching benefits to needs, laying a foundation of agreement, handling any objections, and then stopping and waiting for customers to start shelling out their money. Both specialists are right and **the correct answer is C.**

120. Neither parts specialist is right. The production date is the date that the vehicle was assembled. The standard location for this information is on a tag affixed to the driver's door by the door latch, or on the driver's door sill plate. **The correct answer is D.**

121. Parts Specialist A is correct and **A is the correct answer.** Vehicle build sheets can be a good source of additional information. Vehicle build sheets include the vehicle's identification number (VIN), color, engine size, transmission type, and axle ratio.

122. **The correct answer is D** and only D. The axle tag or stamping is where axle ratio information would be found.

123. Neither specialist is correct. The paint code is not part of the vehicle identification number (VIN). It is located on a separate trim tag, called the service parts identification label, which is mounted under the hood or on the driver's door of some vehicles. **The correct answer is D.**

124. Parts Specialist A is right; product additions can be found in the table of contents of any parts catalog. Parts Specialist B is wrong. The table of contents and index will show the catalog sequence. It will be quite evident if trucks are listed with passenger cars, or if imports are integrated or listed separately. Product additions, possibly resulting from the consolidation of one or more formerly separate catalogs, can be quickly spotted. Instant recognition of several other catalog entities can also be found. These include numerical listings, buyer's guides, progressive size listings, illustrations, installation instructions, competitive interchanges, and more. **The correct answer is A.**

125. Parts Specialist A is wrong, but Specialist B is right. The table of contents and index will show the catalog sequence. It will be quite evident if trucks are listed with passenger cars, or if imports are integrated or listed separately. Product additions, possibly resulting from the consolidation of one or more formerly separate catalogs, can be quickly spotted. Instant recognition of several other catalog entities can also be found. These include numerical listings, buyer's guides, progressive size listings, illustrations, installation instructions, competitive interchanges, and more. **The correct answer is B.**

126. **The correct answer is D** because both specialists are wrong. Bulletins are the primary tool for conducting catalog maintenance, which involves continually updating and revising the catalog racks and making sure counter personnel are using the most up-to-date information from those catalogs.

127. Parts Specialist A is right. It is very important that parts specialists use the most up-to-date information. Parts Specialist B is wrong and is avoiding an important responsibility. Bulletins are the primary tool for conducting catalog maintenance, which involves continually updating and revising the catalog racks and making sure counter personnel are using the most up-to-date information from those catalogs. **The correct answer is A.**

128. As with any delivery, an order for a shipment must be carefully picked, packaged, and documented. A packing list or slip must be included with each shipment. This list contains a detailed description of the items included in the shipment. A separate packing list can be placed in each part of a multiple-part shipment, but often a single comprehensive list is packaged with or secured to the first part of a multiple-part shipment. The only correct statement of the answer choices is D and **D is the correct answer.**

129. A packing list or slip must be included with each shipment. This list contains a detailed description of the items included in the shipment. A separate packing list can be placed in each part of a multiple-part shipment, but often a single comprehensive list is packaged with or secured to the first part of a multiple-part shipment. Both specialists are right and **the correct answer is C.**

130. **The correct answer is C.** Physical inventory should be done once a year. Physical inventory is when all stock is pulled off of the shelves and checked to determine what is actually in the store or shop versus what is in the computer. This is one way to keep the computer system up-to-date and can help you to give the customer faster and more accurate service.

131. Physical inventory should be done once a year. Physical inventory is when all stock is pulled off of the shelves and checked to determine what is actually in the store or shop versus what is in the computer. This is one way to keep the computer system up-to-date and can help you to give the customer faster and more accurate service. Both parts specialists are right and **the correct answer is C.**

132. After the physical inventory has been completed, the parts specialist should report all discrepancies. **The correct answer is C.** The discrepancies are reported by filling out an inventory discrepancy form. After discrepancies have been entered into the computer system, they should be kept and filed for future reference.

133. Parts Specialist A is right. All of the discrepancy forms should be kept and filed. Parts Specialist B is also right. After a discrepancy form has been filled out, it must be corrected in the computer system. **The correct answer is C.**

134. Stock should be rotated every time the shelves are stocked. Stock rotation is moving the older stock in front of the newer stock. This prevents the older stock from sitting at the back of the shelf and eventually getting an aged appearance. Rotating the product so that the labels are toward the front of the shelf is known as facing. Rearranging the shelves for a neater appearance is done whenever necessary to increase appeal to the customer. **The correct answer is B.**

135. Stock rotation prevents older stock from sitting at the back of the shelf and eventually having an aged look. **The correct answer is A.**

136. Parts Specialist A is wrong. A special order is not the weekly stock order. A special order is placed whenever a customer purchases an item not kept in stock. Parts Specialist B is right. A part that repeatedly appears on the special order list could possibly become a part that is stocked in the store. **The correct answer is A.**

137. Parts Specialist A is right. A special order is made for a part that is not stocked in that store. Parts Specialist B is wrong in saying that some parts will always be special order parts. A part that repeatedly appears on lost sales or special order reports should be considered for stocking status. **A is the correct answer.**

138. **The correct answer is C** because both parts specialists are right. The parts counterperson should be careful when handling cores and accepting core returns from customers. A core charge is a charge that is added when the customer buys a remanufactured or reconditioned part. Core chargers are refunded to the customer when the old defective but rebuildable part is returned.

139. A core charge is a charge that is added when the customer buys a remanufactured or reconditioned part. Core chargers are refunded to the customer when the old defective but rebuildable part is returned. To ensure that customers return their cores, a parts specialist should always require a core charge. In this question all of the parts listed may be available for remanufacture, except spark plugs. Often brake shoes, brake calipers, and water pumps are sold as rebuilt units. **The correct answer is D.**

140. A warranty-return part should be kept off the shelf, not to be sold again. Additional forms must usually be completed for warranty-return parts. This allows the manufacturers to determine what is wrong with the part and to credit the store for the return. A warranty-return part should never be placed back in the inventory to be sold to another customer. **The correct answer is D.**

141. Parts Specialist A is right. Additional forms must usually be completed for warranty-return parts. This allows the manufacturers to determine what is wrong with the part and to credit the store for the return. A warranty-return part should never be placed back in the inventory to be sold to another customer. Parts Specialist B is wrong; a brake shoe may have a core charge. **The correct answer is A.**

142. A pair refers to 2. **The correct answer is B.**

143. Each refers to one (1). **The correct answer is D.**

144. An additional freight charge could be added to an emergency order. Freight charges may be added to emergency order parts to cover the cost of special transportation to the store. **The correct answer is C.**

145. Parts Specialist A is right and B is wrong. Freight charges may be added to emergency order parts to cover the cost of special transportation to the store. It is not uncommon to add a long-distance phone charge for special phone orders. Some parts may have a restocking fee that is charged to the customer when the parts are returned for a refund. **The correct answer is A.**

146. The main benefit of using a lost sales report is to keep track of parts that are selling and those that are not selling. **The correct answer is C.** Using lost sales reports allow you to spot trends in the aftermarket parts business. These reports will show new part numbers that the store does not stock as well as older part numbers that are increasingly in demand. Some part numbers are assured of almost instant demand, such as those found in some of the newer downside models. Others parts will grow in demand, but somewhat more slowly. The storeowner or buyer must analyze not only the part numbers that are selling, but also those that the store does not stock.

147. **The correct answer is D.** The type of report shown in the figure is a lost sales report.

148. In this question, worn valve lock grooves are being discussed. Parts Specialist A says worn valve lock grooves may cause the valve locks to fly out of place with the engine running, resulting in severe engine damage. Specialist A is right. Parts Specialist B is wrong. Worn valve lock grooves probably will not cause a clicking noise with the engine idling. They may cause the spring to leave its retainer. **The correct answer is A.**

149. The parts specialists are discussing the results of a cylinder balance test. Cylinder number three provides very little drop in rpm when disabled. Parts Specialist A says the ignition system may be misfiring on the number three cylinder. Parts Specialist B says the engine may have an intake manifold vacuum leak. They both are right and **the correct answer is C.** If the cylinder is working normally, a noticeable decrease in engine speed occurs when the cylinder misfires. If there is very little decrease when the analyzer causes a cylinder to misfire, the cylinder is not contributing to engine power. Under this condition the engine compression, ignition system, and fuel system should be checked to locate the cause of the problem. An intake manifold vacuum leak may cause a cylinder misfire with the engine idling or operating at low speed. If this problem exists, the misfire will disappear at a higher speed when the manifold vacuum decreases. When all cylinders provide the specified drop in engine speed, all are contributing equally to the engine power.

150. **The correct answer is D.** A collapsed upper radiator hose is caused by a decrease in pressure. This normally occurs when the temperature of the system decreases. Some radiator hoses contain a wire coil inside them to prevent hose collapse. Parts Specialist A is wrong. A defective pressure release valve in the radiator cap would cause an increase in pressure and may cause a hose or hose connection to blow. Parts Specialist B is also wrong. A restriction in the hose between the radiator filler neck and the recovery reservoir would also cause pressure to increase.

151. A thermostat that is stuck open would tend to allow the coolant circulating through the engine to be cooler than normal. Excessive high coolant levels in the recovery reservoir are caused by those things that cause high temperature. As temperature increases, pressure increases. The higher pressures push more fluid into the reservoir. **The correct answer is B.** All of the answer choices would cause high temperatures except a stuck thermostat.

152. The wastegate of a turbocharger is designed to prevent high output pressures and to regulate the maximum output from the turbocharger. Of the answer choices for this question only one would cause reduced turbocharger output. This would be choice B. A stuck open wastegate would always allow some of the pressure from the turbocharger to bleed off and the pressure would always be low. The other choices would probably cause an over boost situation. **The correct answer is B.**

153. **The correct answer is B.** If the fuel pump pressure is less than specified, the air/fuel ratio would be leaner than normal, not too rich. The rest of the statements are true. The fuel system should be depressurized prior to the fuel pump pressure test, unless the system is equipped with a Schrader valve. The pressure gauge should be connected to the Schrader valve on the fuel rail or into the main fuel inlet at the throttle body. Many things can cause lower than specified fuel pressure; one of these may be a restricted fuel filter.

154. The meter shown is connected across the ground circuit of the solenoid. This hookup would display the voltage drop across the ground. In an ideal world, this reading should be zero. Neither specialist is right and **the correct answer is D.**

155. Low maximum secondary voltage is probably caused by a problem in the primary circuit. The efficiency of the primary has a direct effect on the efficiency of the secondary. Only one of the answer choices would cause decreased output or poor performance of the primary circuit. That would be choice B: a low primary input. Low input would definitely result in low output; therefore **answer B is the correct answer.** The other answer choices would not result in low maximum secondary coil voltage. Low primary resistance would provide for an increased output from the primary, which in turn would increase the output from the secondary. Wide spark plug gaps and open spark plug wires would not affect the output from the coil. They would, however, require high firing voltages.

156. **The correct answer is D** because both parts specialists are wrong. The backpressure reading is taken with a warm engine at 2,500 rpm. A restriction is indicated if the pressure exceeds 1.75 psi (12 kPa). The temperature test is taken after the engine has been running at 2,500 rpm for three minutes. If the carbon monoxide (CO) exceeds 0.3 percent, the oxygen (O_2) exceeds 0.4 percent, and the hydrocarbon (HC) content falls between 120 and 400 parts per million (ppm), the catalytic converter is probably defective.

157. Both parts specialists are wrong. The positive crankcase ventilation (PCV) valve moves toward the open (not closed) position when the throttle is opened from idle to half throttle. The PCV valve is moved toward the open (again not closed) position by spring tension. **The correct answer is D.**

158. When diagnosing a PCV system, the first step is to check all the engine gaskets for signs of oil leaks. Be sure the oil filler cap fits and seals properly. Check the clean air hose and the PCV hose for cracks, deterioration, loose connections, and restrictions. Check the PCV clean air filter for contamination and replace if necessary. If there is evidence of oil in the air cleaner, check the PCV valve and hose for restriction. Parts Specialist A says the positive crankcase ventilation (PCV) valve clean air filter in the air cleaner may be plugged. If this were the case, the system would not be able to draw in fresh air and this may cause oil dilution. Therefore A is wrong. Parts Specialist B says the hose from the positive crankcase ventilation (PCV) valve to the intake manifold may be severely restricted. This would cause crankcase pressures to escape through the PCV air filter and collect in the air cleaner housing. B is correct. **The correct answer is B.**

159. **The correct answer is A.** When diagnosing a catalytic converter with a digital pyrometer, if the converter is operating properly, the converter outlet should be 100°F (38°C) hotter than the inlet. This shows the converter is converting and working well.

160. **The correct answer is A.** Repeated extension housing seal failure may be caused by a scored driveshaft yoke. The other answer choices would have little to no effect on the wear of the seal.

161. **The correct answer is D.** The second speed gear dog teeth and blocking ring teeth are badly worn. This problem may cause the transaxle to jump out of second gear. Once a gear has been selected, the synchronizer assembly keeps the driven gear locked to its shaft. This in turn keeps the driven gear in mesh with the drive gear. If the synchronizers are defective or worn, the gear can unlock from the shaft and the result would be that the transaxle will not stay in the selected gear. There are other causes for this problem, such as worn or loose engine mounts.

162. Automatic transmission fluid (ATF) is pink or red. If the fluid is dark brown or blackish and has a burned odor, the fluid has been overheated, possibly from burned clutches or bands. A milky-colored fluid is most likely caused by coolant contamination from a leaking transmission cooler. If silvery metal particles are found in the fluid, it is an indication of damaged transmission components. If the dipstick feels sticky and is difficult to wipe clean, the fluid contains varnish, an indication that ATF and filter changes have been neglected. In this question the fluid is a dark brown color and has a burned smell. Parts Specialist A is wrong. This problem may be caused by burned clutches or bands, not by a worn front planetary sungear. Parts Specialist B is right. **The correct answer is B.**

163. **The correct answer is B.** An improper band adjustment may cause transmission slipping in some gears. A and B are wrong; the band has nothing to do with the speed of shifting. This is controlled by the valve body, solenoids, governor, and other shift control mechanisms. Since a band is used for one or two gears, a faulty band would not affect all gears. If the transmission slipped in only one gear and that gear involved the engagement of the band, then a faulty band would cause the problem.

164. The question refers to the diagnosis of driveshaft and universal joint (U-joint) in rear-wheel-drive (RWD) vehicles. Parts Specialist A is wrong in saying that a worn U-joint may cause a squeaking noise that decreases in relation to vehicle speed. A worn joint will typically cause a clanking noise when the transmission shifts gear or when the vehicle is accelerating or decelerating. Parts Specialist B is right. A heavy vibration that only occurs during acceleration may be caused by a worn centering ball and socket on a double Cardan U-joint. **The correct answer is B.**

165. The item in the drawing is a thermal expansion valve. **The correct answer is B.**

166. Neither parts specialist is right and **the correct answer is D.** Flexible brake hoses allow for movement between the suspension and the chassis. These hoses should be inspected for cracks, leaks, twists, bulges, loose supports, and internal restrictions. Each time a brake hose is removed, the sealing washer on the male end should be replaced. When a brake hose is installed, always install and tighten the male end first.

167. Both parts specialists are right. The backing plate should be inspected for distortion, cracks, rust damage, and wear in the brake shoe contact areas. A distorted backing plate may cause brake grabbing. Check the anchor bolt for looseness that may result in brake chatter. **The correct answer is C.**

168. When two tapered roller bearings are mounted in the front or rear wheel hub, a typical bearing adjustment procedure is as follows: (1) tighten the bearing adjustment nut to 17 to 25 ft.-lbs. (23 to 34 N-m), (2) back off the adjustment nut one-half turn, (3) tighten the bearing adjustment nut to 10 to 15 ft-lbs. (13.6 to 20.3 N-m), and (4) install the nut retainer and cotter key. **The correct answer is C.** After servicing and lubricating the rear wheel bearings in a front-wheel-drive car, the bearing adjusting nut is tightened to 20 ft.-lbs. (27 N-m) and loosened one-half turn. The next step in the bearing adjusting procedure is to tighten the adjusting nut to 10 to 15 ft.-lbs. (13.6 to 20.3 N-m).

169. All of the following statements about caster adjustment are true. The front wheels are turned 20 degrees outward and then 20 degrees inward to read the caster angle. The brakes must be applied with a brake pedal jack before reading the caster angle. The front suspension should be jounced several times before reading the caster angle. Answer A is not true. Caster is not measured with the front wheels straight ahead. **The correct answer is A** because it is wrong.

170. The question is focused on a receiver/dryer that is located between the condenser outlet and the evaporator inlet. Parts Specialist A is right in saying the receiver/dryer should be changed if the outlet is colder than the inlet. This would indicate some kind of restriction at the outlet. Parts Specialist B is wrong. Using the sight glass can help in diagnosing the A/C system. Grayish blue particles in the sight glass are an indication that the desiccant in the receiver/dryer has disintegrated and is circulating through the refrigeration system. This condition also requires receiver/dryer replacement. If the refrigerant in the sight glass is red or yellow, leak-detecting dye has been added to the refrigeration system. This condition does not require corrective action. All of the refrigerant must be recovered from the system before receiver/dryer removal. **The correct answer is A.**

171. Both parts specialists are right. When the discharge air temperature is higher than specified in an air conditioning (A/C) heater system with a coolant flow control valve when the system is operating in the maximum A/C mode, the coolant flow control valve may be stuck in the open position, or the coolant flow control valve vacuum hose may be disconnected or have a leak. **The correct answer is C.**

172. Both parts specialists are wrong. When measuring resistance with an ohmmeter, the circuit or component must be disconnected from power. If this is not done, the circuit's current will damage the meter. Also, when testing a spark plug wire with a 20,000 Ohm resistance, you should use the x1K scale not the ×100 scale. **The correct answer is D.**

173. Both parts specialists are wrong and **the correct answer is D.** Some wiper motors contain a series field coil, a shunt field coil, and a relay. When the wiper switch is turned on, the relay winding is grounded through one set of switch contacts. This action closes the relay contacts, and current is supplied through these contacts to the series field coil and armature. Under this condition the wiper motor starts turning. If the wiper switch is in the high-speed condition, the shunt coil is not grounded and the motor turns at high speed. When the wiper switch is in the low-speed position, the shunt coil is grounded through the second set of wiper switch contacts. Under this condition current flows through the shunt coil and the wiper switch to ground. Current flow through the shunt coil creates a strong magnetic field that induces more opposing voltage in the armature windings. This opposing voltage in the armature windings reduces current flow through the series coil and armature windings to slow the armature. If the wiper motor fails to park, or parks in the wrong position, the parking switch or cam probably is defective.

174. This is an except-type question. All of the statements relating to motor oil in this question are true except for choice B. Motor oil is a high volume, but is not a high profit item. **The correct answer is B.** Customers are very sensitive to motor oil prices. Changing motor oil is the most popular DIY product. Stocking motor oil is a necessity.

175. **The correct answer is B** because this number is not an alphanumeric. The rest are.

176. Both specialists are right and **the correct answer is C.** Gasoline engines and diesel engines are internal combustion engines.

177. **The correct answer is C.** The camshaft rotates.

178. **The correct answer is D.** Flywheel runout is measured with the use of a micrometer. Inspect the flywheel for scoring and cracks in the clutch contact area. Minor score marks and ridges may be removed by resurfacing the flywheel. Mount a dial indicator on the engine flywheel housing, and position the dial indicator stem against the clutch contact area on the flywheel. Rotate the flywheel to measure the flywheel runout. If the flywheel runout exceeds specifications it must be replaced.

179. Blue exhaust is an indication of excessive oil consumption. **The correct answer is B.** A sound engine would be nearly invisible. The exhaust from an engine that has a rich air/fuel mixture would have a blackish exhaust. Coolant leaking into the combustion chamber would cause a whitish exhaust.

180. **The correct answer is D.** The tool used to measure power steering pump rotors is a micrometer.

181. A thermostat that is stuck open reduces coolant temperature and does not cause high coolant level in the recovery tank. The radiator cap should be inspected for a damaged sealing gasket or vacuum valve. If the pressure cap is damaged, the engine will overheat and coolant will be lost to the coolant recovery system. Under this condition the coolant recovery tank becomes overfilled. If the cap valve is sticking, a vacuum may occur in the cooling system after the engine is shut off and the coolant temperature decreases. This vacuum may cause collapsed cooling system hoses. A pressure tester may be used to test the pressure cap and the entire cooling system. Based on the answer choices, the LEAST likely problem if the thermostat is stuck in the open position would be overheating. **The correct answer is B.**

182. **The correct answer is C.** A missing air filter and a clogged fuel filter will affect engine performance.

183. Worn turbocharger seals may be noted by blue smoke from the exhaust, due to the oil in the exhaust. **The correct answer is D.**

184. Mechanical or electric fuel pumps should be tested for pressure and flow or volume. A pressure gauge is connected in series in the fuel inlet line on the carburetor or throttle body assembly on throttle body injection (TBI) engines. On most port fuel injection engines the pressure gauge is connected to the Schrader valve on the fuel rail. **The correct answer is D.**

185. All of the answer choices are part of ignition system, except B, the battery. **The correct answer is B.** The ignition system is a major component of a gasoline engine powered vehicle. It consists of the battery, ignition coil, low voltage (primary) wiring, ignition start/run switch, distributor (some vehicles are equipped with a DIS, or distributorless ignition system), high-tension (secondary) wiring, and spark plugs. The ignition system provides the right spark at the right time to the right cylinder to ignite the air/fuel mixture in the combustion chamber.

186. The statement that is not true is C. The meter is connected to measure the resistance across a diode. A good diode will have high resistance when the meter is connected in one direction. If the leads of the meter are reversed, the meter should show a low resistance reading. **The correct answer is C.**

187. A typical exhaust system has the following components: exhaust manifold and gasket; exhaust pipe, seal, and connector pipe; intermediate pipe(s); catalytic converter; resonator and/or muffler; tailpipe; and hardware items including heat shields, clamps, gaskets, and hangers. Of the answer choices, D is the item not found on all vehicles. **The correct answer is D.**

188. When checking a PCV valve, air should pass freely in one direction but not the other. **The correct answer is B.** Connect a length of hose to the inlet side of the PCV valve and blow air through the valve with your mouth while holding your finger near the valve outlet. Air should pass freely through the valve. If air does not pass freely through the valve, replace the valve. Connect a length of hose to the outlet side of the PCV valve and try to blow back through the valve. It should be difficult to blow air through the PCV valve in this direction. When air passes easily through the valve, replace the valve.

189. The evaporative emission controls (EEC) system draws vapors from the fuel tank and introduces them into the intake air stream. **The correct answer is D.**

190. The exhaust gas recirculation (EGR) system introduces exhaust gases into the intake air to reduce the formation of oxides of nitrogen in the combustion chamber. The EGR valve uses ported vacuum as its primary vacuum source. Ported vacuum is used because EGR is not needed at idle. On many late-model vehicles, a positive backpressure or negative backpressure EGR valve is often used. This type of valve requires a certain level of backpressure in the exhaust before it will open when vacuum is applied. The vacuum control plumbing to the EGR valve usually includes a temperature vacuum switch (TVS) or solenoid to block or bleed vacuum until the engine warms up. On some late-model vehicles with computerized engine controls, the computer actuates the solenoid to further modify the opening of the EGR valve. Only Specialist B is right and **the correct answer is B.**

191. When the engine is not running, or during a backfire, the positive crankcase ventilation (PCV) valve plunger is seated in the housing, and the passage through the valve is closed. If the engine is idling, the high intake manifold vacuum holds the tapered valve plunger nearly closed. When the throttle is opened and the manifold vacuum decreases, the tapered plunger gradually moves downward to provide more PCV valve opening. If the PCV valve is stuck closed, the air/fuel ratio is richer and hydrocarbon (HC) and carbon monoxide (CO) emissions are higher. A PCV valve stuck in the open position may cause a leaner air/fuel ratio and increased engine idle speed. If the inside

of the air cleaner is contaminated with engine oil, the engine may have excessive blowby, or the PCV valve and connecting hose may be restricted. When the PCV clean air filter in the air cleaner is plugged, higher vacuum is built up in the engine. This may cause damage to the engine gaskets and pull in dirt particles that could cause damage to the engine. All of the answer choices for this question are true, except D, and **D is the correct answer** to this except-type question.

192. Both parts specialists are wrong and **the correct answer is D.** When the clutch is engaged, the spring pressure plate forces the disc hard against the face of the flywheel which, when the engine is running, causes the flywheel, disc, and pressure plate to rotate together. When the clutch is disengaged, the pressure plate is pulled away from the disc, allowing the flywheel and pressure plate to turn without rotating the disc.

193. Automatic transmission fluid (ATF) is pink or red. If the fluid is dark brown or blackish and has a burned odor, the fluid has been overheated, possibly from burned clutches or bands. A milky-colored fluid is most likely caused by coolant contamination from a leaking transmission cooler. If silvery metal particles are found in the fluid, it is an indication of damaged transmission components. If the dipstick feels sticky and is difficult to wipe clean, the fluid contains varnish, an indication that ATF and filter changes have been neglected. **The correct answer is C.**

194. The driveshaft is nothing more than an extension of the transmission output shaft. That is, the driveshaft, which is usually made from seamless steel tubing, transfers engine torque from the transmission to the rear-driving axle on a rear-wheel-drive vehicle. The yokes, which are either welded or pressed on the shaft, provide a means of connecting two or more shafts together. At the present time, a limited number of vehicles are equipped with fiber composite-reinforced fiberglass, graphite, and aluminum driveshafts. The advantages of the fiber composite driveshaft, other than weight reduction and torsional strength, are fatigue resistance, easier and better balance, and reduced interference from shock loading and torsional problems. The U-joint allows two rotating shafts to operate at a slight angle to each other. **The correct answer is D.**

195. Parts Specialist A is right. Constant velocity (CV) joints should be replaced as a complete assembly. Parts Specialist B is wrong. When reassembling the joint always install all the grease in the joint that is provided in the repair kit. **The correct answer is A.**

196. Specialist B is correct. This setup is used to measure the thickness of a rotor. This thickness is compared to the minimum thickness specs. If the rotor is less than specifications, the rotor should be discarded. Rotor distortion is checked with a dial indicator. **The correct answer is B.**

197. **The correct answer is B.** The front brakes do as much as 80 percent of the work in stopping a car. Older front/rear split hydraulic systems have given way to dual diagonal split systems. These systems increase stopping power and provide 50 percent of the braking capacity in case of failure in either of the two hydraulic systems.

198. Both parts specialists are wrong and **D is the correct answer.** Although some USC and metric head sizes are very close, never use a metric wrench or socket for USC bolts, or vice versa. Tool slippage may cause injury or damage the bolt head. Bolt diameter is the measurement across the major diameter of the threaded area or across the bolt shank. The thread pitch of a bolt in the English system is determined by the number of threads there are in one inch of threaded bolt length and is expressed in number of threads per inch. The thread pitch in the metric system is determined by the distance, in millimeters, between two adjacent threads. To check the thread pitch of a bolt or stud, a thread pitch gauge is used. Gauges are available in both English and metric dimensions.

199. Both parts specialists are right. Bolt diameter is the measurement across the major diameter of the threaded area or across the bolt shank. The thread pitch of a bolt in the English system is determined by the number of threads there are in one inch of threaded bolt length and is expressed in number of threads per inch. The thread pitch in the metric system is determined by the distance in millimeters between two adjacent threads. To check the thread pitch of a bolt or stud, a thread pitch gauge is used. Gauges are available in both English and metric dimensions. Bolt length is the distance measured from the bottom of the head to the tip of the bolt. The bolt's tensile strength, or grade, is the amount of stress or stretch it is able to withstand. The type of bolt material and the diameter of the bolt determine its tensile strength. In the English system, the tensile strength of a bolt is identified by the number of radial lines (grade marks) on the bolt head. More lines mean higher tensile strength. In the metric system, tensile strength of a bolt or stud can be identified by a property class number on the bolt head. The higher the number, the greater the tensile strength. **The correct answer is C.**

200. The tool shown is plastigage and is used to measure the clearance between the bearing and shaft. A strip of thread is placed between the shaft and the bearing. How much the thread spreads depends on the clearance. The less clearance, the more the thread will spread. **The correct answer is B.**

Glossary

Accessories Parts that add to the appearance or performance of a vehicle.

Accounts receivable Money due from customers.

Accumulator Part of an air conditioning system which contains a desiccant that absorbs moisture from the refrigerant.

Active accounts Current customers who make frequent purchases.

Active stock Merchandise in the store that is readily available for sale to customers.

Air filter Part of an engine intake system which cleans dirt and dust from the air before it enters the engine.

Air Injection Reaction (AIR) An emissions control system which pumps air into the exhaust system to burn hydrocarbons, carbon monoxide, and oxides of nitrogen coming from the engine.

Air pump Part of the Air Injection Reaction (AIR) system which forces extra air to mix with the exhaust gases. This action causes the continued burning of any hydrocarbons and carbon monoxide remaining in the exhaust.

Alphanumeric A numbering system consisting of a combination of letters and numbers. They are placed in order starting from the left digit and working across to the right. This system is commonly used in parts catalogs and price sheets.

Alternating current (AC) An electrical current which flows alternately in two directions, forward and backward. It is produced by some form of mechanical device or motion, such as an alternator. Alternating current cannot be used to charge a battery. It must first be converted (rectified) to direct current for battery charging.

Alternator The electrical device driven by the engine which recharges the battery. It produces alternating current using rotating field coils inside stationary stator windings.

Anti-lock brakes A brake system which operates by pulsing the pressure to the wheel and caliper cylinders to prevent the lockup of any one wheel, thus preventing a skid.

Automatic transmission A transmission which automatically selects the correct gear ratio needed for the driving conditions. The driver has only to select the direction of travel desired.

Automotive aftermarket The distribution and sale of automotive replacement products.

Back order Merchandise ordered from a supplier but not shipped due to the supplier being out of stock.

Ball joint Part of the steering system which connects the spindle to the control arm. It allows the steering knuckle to turn right and left as well as permitting the control arm to move up and down.

Band A flexible flat piece inside an automatic transmission which holds parts of the gearset to create the correct gear ratio.

Battery A device within the electrical system which stores voltage until it is needed for vehicle operation.

Bearing A term used for ball bearing; an antifriction device having an inner and outer race with one or more rows of hardened steel balls between them.

Bill of lading A shipping document acknowledging receipt of goods and stating terms of delivery.

Blend door Part of the air conditioning/ventilation system which controls the ratio of incoming heated air and fresh air. This ratio is adjusted to control the temperature of the passenger compartment.

Blowby The unburned fuel and combustion byproducts which leak past the piston rings and into the crankcase.

Brake pads The friction elements of a caliper-type brake system. They are usually forced toward the rotor by hydraulic pressure.

Brake shoes The friction elements of a drum-type brake system. They are usually expanded outward by hydraulic pressure to contact the inside diameter of the brake drum.

Brake spring pliers Pliers used in the disassembly of brake drum components.

Break point Where cost of shipping by a particular method changes significantly because of size or weight classifications. For example, parcel post shipments cannot exceed 40 pounds.

Bushing A smooth cylinder used to reduce friction and to guide the motion of the parts.

Camshaft The shaft in an engine which causes the valves to open and close at the correct times.

Capacitor An electrical device used to smooth out changes in voltage levels.

Carbon dioxide A colorless, odorless, incombustible gas which is exhaled by humans and required by plant life. In the automotive emissions system, carbon monoxide is mixed with air (oxygen) to form harmless carbon dioxide.

Carbon monoxide A colorless, odorless, toxic gas formed by internal combustion engines.

Carburetor A device used on some engines to mix the fuel and air in the correct ratio for efficient combustion.

Cash discount A discount given for the prompt payment of a bill.

Cataloging The process of looking up the needed parts in the parts catalog.

Catalytic converter A metal canister mounted in the exhaust system containing metals that convert harmful exhaust gases into safer gases. A catalytic converter speeds up the reaction but is not consumed in the chemical reaction.

Caustic A compound that is able to burn, corrode, or eat away another compound.

Celsius The metric unit of temperature measurement.

Check valve A valve which allows something to move or flow in only one direction.

Circuit breaker An electrical device which protects the circuit by interrupting the current flow when it exceeds the rated capacity. It may be reset manually or automatically.

Closed loop An operating state in a computer-controlled engine in which the computer is controlling engine operation based upon information created by the sensors.

Closing a sale Getting a commitment to buy from a customer.

Clutch (1) The part of a driveline system which is used to interrupt the power flow between the engine and the transmission. (2) A part mounted on the air conditioning (AC) compressor used for temperature control of the AC system.

Clutch-release lever The part of a manual transmission system which operates the clutch-release (throwout) bearing.

Coil spring A spring which is wound into a spiral shape. Coil springs are commonly used on automotive suspension systems.

Combustion chamber The area inside the cylinder head and block where the burning of the fuel takes place.

Compression The act of forcing something together. In an automotive engine, the air and fuel are compressed in the cylinders to create a larger explosion and thus more power upon ignition.

Compressor The part of the air conditioning system which compresses the refrigerant vapor and pumps the refrigerant.

Computer terminal A keyboard system that permits a parts specialist, counter person, or operator to input invoices or obtain information from a computer usually located in a warehouse or distribution offices.

Condenser The part of the air conditioning system which cools the hot vapor and converts it to a liquid. The condenser is usually mounted in front of or on top of a vehicle for better airflow.

Constant velocity (CV) joint Part of the drive train which allows for changes in the angle of a driveshaft or half shaft.

Control arm Suspension parts which control spring action and the direction of travel of the axle as it reacts to driving conditions.

Coolant A fluid used for cooling, usually consisting of a blend of water and antifreeze.

Core Items such as starters, alternators, carburetors, and brake shoes accepted in exchange for remanufactured items.

Core charge A charge which is added when the customer buys a remanufactured part. Core charges are refunded to the customer when the core is returned.

Correction bulletin A bulletin which corrects catalog errors due to printing errors or inaccurately assigned part numbers.

Counter cards Advertising or display placards, usually with an easel back, placed on the counter.

Countershaft A shaft used in transmissions to transfer the motion from the input shaft to the output shaft.

Crankcase The lower part of an engine.

Crankshaft The shaft in an engine which converts the reciprocating piston motion into rotary motion for the driven device.

Credit memo The record of an amount paid to a customer, usually for the return of a purchased item or a core.

Crossover pipe The pipe which connects the two exhaust pipes of a dual exhaust system.

Cubic inch The volume equal to a cube with one-inch sides. The term is commonly used to describe engine displacement.

Customer relations A description of how a salesperson interacts with the customer.

Cylinder head The removable part covering the top of the engine cylinders. It seals the combustion chamber and usually contains the valves and spark plugs.

Dating Payments for merchandise extended to 30, 60, or 90 days without loss of cash discounts.

Dealers The jobber's wholesale customers, such as service stations, garages, and car dealers, who install parts in their consumers' vehicles.

Demand Items Items such as water pumps, bearings, clutches, and remanufactured parts that a customer needs for a specific vehicle.

Deposit A specified sum of money a customer leaves with the jobber for special orders or to guarantee the return of a core.

Desiccant Any substance used to absorb moisture. Desiccant is used in an air conditioning system to keep moisture from forming corrosive compounds.

Diameter The distance straight across a circular figure; the largest measurement which can be taken across a circular object.

Differential The set of gears which transmits power from the driveshaft to the wheels and allows the drive wheels to turn at different speeds for cornering.

Direct current (DC) An electrical current which flows in only one direction. It is usually created from a chemical source, such as a battery, and can be used to recharge a battery.

Direct mail Advertising literature sent to customers through the mail.

Discount A deduction from the stated price, usually a percentage off the printed price sheet.

Display A special grouping of sale items or new product lines with point-of-sale materials designed to attract customers' attention.

Distributor (1) The part of the ignition system which directs the secondary voltage (spark) to the correct spark plug. (2) A large volume stocking parts house that sells to wholesalers.

Do-it-yourselfer Someone who performs service and repair on his or her own vehicle.

Drag link The rod which connects the steering box to the steering knuckle on a straight axle front.

Driveshaft The shaft which connects the transmission to the differential.

Dual account A classification of a customer according to the purposes of the purchases. For example, a dual account customer can be one who both uses and sells parts, or a customer who uses parts for industrial and automotive purposes. The term is used in wholesaler policies that affect discounts, taxes, or credit.

Dual exhaust An exhaust system which uses separate exhaust pipes for each bank of cylinders.

Early Fuel Evaporization (EFE) An emissions system used to warm the air going into a cold engine to improve driveability and reduce hydrocarbon and carbon monoxide emissions.

Emergency order An order placed out of the normal cycle of stock orders, usually when unexpected or emergency parts are needed.

End cap A display fixture located at the end of gondolas.

Engine block The main body of an engine. The block contains the cylinders and carries the accessories.

Engine coolant The solution used in an engine to carry heat away from the cylinders. It is usually a 50/50 mixture of ethylene glycol and water.

Evaporator The part of an air conditioning system which is mounted inside the passenger compartment. It contains low-pressure refrigerant liquid and vapor to remove heat from the air forced through and around it.

Excise tax Taxes on products paid by manufacturers or sellers usually included in the price of an item, but often added, as in the case of tires.

Exhaust Gas Recirculation (EGR) An emissions device which meters some exhaust gas into the intake manifold to reduce the combustion chamber temperature and to prevent the formation of oxides of nitrogen.

Exhaust manifold The part of the exhaust system which bolts directly to the cylinder heads. Exhaust gases leaving the cylinder are routed through the cylinder head to the exhaust manifold.

Extra dating A discount is made available for purchased items that are delivered at a given date and marked payable in a given period, such as 30, 60, or 90 days. It is used to spread payment without loss of the cash discount.

Facing Rotating a product so its labels are facing toward the front of the shelf. Facing improves shelf appearance and customer appeal.

Fahrenheit The temperature unit used in the English system of measurement.

Fastback A vehicle body style which has a fixed rear glass. Access to the luggage compartment is from the inside of the vehicle.

Fixtures The furnishings in a store such as gondolas, endcaps, display cases, racks, shelving, and counters.

Flare A type of tubing connection which requires that the end of the tubing be expanded at a 37 or 45 degree angle.

Flasher The part of the lighting electrical system which causes the turn signals and hazard lights to blink.

Fleet A number of vehicles operated by one owner or company.

Franchise A specified sum of money an individual must pay for the privilege of owning and operating one of a chain of retail operations.

Freight bill A bill that accompanies goods shipped describing the contents, weight, point of origin, shipper, and giving the transportation charges.

Freight charge A charge added to special order parts to cover transportation to the store.

Fuel injector The fuel system component which sprays fuel into the intake manifold on vehicles not equipped with a carburetor.

Fuel pump The fuel system component which moves the fuel from the tank to the carburetor or the fuel injection unit.

Fuse An electrical device which protects the circuit by interrupting the current flow when the flow exceeds the fuse's rated capacity. Fuses are constructed of a conductor encased in plastic or glass and must be replaced when blown.

Fusible link An electrical device which protects the circuit by interrupting the current flow when the flow exceeds the rated capacity. A fusible link is constructed of a wire surrounded by a special insulation and must be replaced when blown.

General application items Products such as oil, polish, antifreeze, and chemicals that apply to all makes of vehicles.

Gondolas Long, shelved display fixtures usually placed back to back to divide stores into aisled trafficways.

Grade (1) A method of classifying bolt strength. The grade of a bolt is indicated by markings on the head. (2) The classification of a product, such as motor oil.

Gross profit The selling price of an item less the cost.

Half shaft One of two shafts which connect a transaxle to the drive wheels.

Handling charge The cost charged to a customer for returning an item to a supplier or manufacturer for repairs or adjustment.

Hatchback A vehicle body style in which the rear glass will lift to allow access to the trunk.

Headers A tubular form of exhaust manifold which is used to increase exhaust gas flow and improve performance.

High-volume Describes a popular item which is sold in large quantities.

Hydrocarbon An organic compound made up of hydrogen and carbon. Most automotive fuels and lubricants contain hydrocarbons, a common source of pollution. As an automotive term, it refers to unburned fuel in the exhaust.

Hydrogen A colorless, highly flammable gas which is the most common in the universe. It is used in the production of methanol, an automotive fuel and fuel additive.

Idle A condition in which the engine is running at a low speed.

Idler arm A part of the steering system which connects the center link to the frame.

Ignition coil A part of the ignition system which produces the electricity needed to create the spark for the spark plugs.

Ignition system The part of the automotive electrical system which creates and delivers the high-voltage spark to the spark plugs.

Impulse item Any item that the customer did not intend to purchase when entering the store.

Inboard A position reference relating to being toward the center, or inside, of the vehicle.

Individually priced The condition of having the price of each part on the display marked for the customer's convenience.

Input shaft The shaft which carries torque into the transmission.

Insert bearings A split circular shell which is inserted between the engine block or connecting rods and the crankshaft.

Intake manifold The part of the intake system which bolts directly to the cylinder heads. The air or air/fuel mixture must pass through the intake manifold before entering the cylinder head and combustion chamber.

Intake valve The valve in the intake port which opens to allow the air/fuel mixture to enter the combustion chamber.

Intercooler A part of the intake system used with a turbocharger to cool the air entering the intake manifold.

Integrated circuit A miniaturized electrical component—consisting of diodes, transistors, resistors, and capacitors—used in electronic circuits.

Interchange list A cross reference for part numbers of identical items from different manufacturers.

Intermediate order An order placed after a regular stock order to replenish stock until the regular stock order shipment is received.

Inventory The goods a store has in its possession for resale.

Inventory control A method of determining amounts of merchandise to order based on supplies on hand and past sales.

Invoice A sales slip used as a record of sales.

Invoice register A system used to maintain a check on the issuing of invoices.

Jobber The owner or operator of an auto parts store usually wholesaling products to volume purchasers such as dealers, fleet owners, and industrial firms, and retailing to do-it-yourselfers.

Journal The part of a shaft which is supported by and comes in contact with a bearing.

Julian A calendar system devised by Julius Caesar which numbers the days consecutively starting with January 1. The Julian calendar is often used in production codes and computer systems.

Lifter A component of the automotive valve train which converts the rotary motion of a camshaft lobe into the reciprocating motion needed to open and close the valves.

Limited-slip differential A differential which uses internal clutch plates to limit the slip and speed difference between the drive wheels. This limiting improves traction on slick surfaces and helps eliminate wheel spinning.

Line The product list of a specific manufacturer.

List price The suggested selling price to the final consumer.

Liter A unit of volume used in the metric system of measurement. A liter is slightly greater than a quart in the English system.

Locator pad A quick-reference chart which will list a part's description, on-hand quantity, cost, and location in part number order.

Lost sale A customer not purchasing a requested item either because the store does not stock the item or because the store stocks brands other than the one requested by the customer.

Machine work Work done in a machine shop. In automotive applications, such work refers to engine block boring, shaft journal turning, brake drum and rotor turning, and cylinder head repair services.

MacPherson strut A suspension system component which combines a lower lateral link with a vertical strut to combine the features of a spindle and a shock absorber.

Main bearing caps The caps which secure the crankshaft to the block.

Maintenance free A term generally associated with a battery that requires no water or charging under normal conditions for the length of the warranty.

Margin Same as gross profit: the cost subtracted from the selling price.

Marketing The total function of moving merchandise from the manufacturer into the hands of the final consumer; including buying, selling, transporting, storing, and advertising.

Mark-up The amount a merchant charges for merchandise above costs to earn a profit.

Material Safety Data Sheet (MSDS) A prepared printed sheet which accompanies chemicals and contains information regarding the proper application along with the safety and environmental concerns.

Media A vehicle used for advertising, such as newspapers, magazines, television, radio, and outdoor posters.

Merchandising The "second effort" of advertising, such as building store displays of advertised items, posting newspaper advertisements and sales brochures, and promoting events through public relations.

Micrometer A precision measuring tool which can measure to dimensions of 0.0001 inch (English) or similar dimensions in the metric system.

Microprocessor A small computer-like device used to process signals from various circuits to achieve a control system that will adapt to operating changes.

Modules Self-contained electronic circuits for computerized systems; usually repaired by plugging in a new unit.

Muffler The part of the exhaust system which is used to reduce exhaust noise.

Net price The cost of an item to a particular purchaser, such as dealer net, jobber net, or user net.

Net profit The merchant's profit after deducting costs of merchandise and all expenses involved in operating the business.

No-return policy A store policy that certain parts cannot be returned after purchase. It is very common on electrical and electronic parts.

Obsolescence When a part is no longer of use, either because of replacement by a superseding part, or due to lack of demand.

OE An abbreviation for original equipment.

OEM An abbreviation for original equipment manufacturer.

Oil filter A device on the engine for removing dirt, carbon, and other impurities from the lubricating oil.

On-hand The quantity of an item that the store has in stock.

Operating capital The money that a merchant needs in everyday operations to buy parts, pay salaries, and meet regular expenses.

Open loop An operating state in a computer-controlled engine in which the computer is controlling engine operation based upon a predetermined program. It is usually in effect until the engine sensors signal that the engine has reached operating temperature.

Orifice tube A part of an air conditioning system which causes the drop in refrigerant pressure that results in the change in temperature necessary for cooling.

OSHA An abbreviation for Occupational Safety and Health Administration. All companies with a specified minimum number of employees must comply with its safety regulations.

Outboard A position reference indicating a state of being toward the outside of the vehicle.

Overage More items received than ordered.

Overhead The costs of operating a business not including purchases of merchandise.

Overhead cam An engine valve train system which has the camshaft positioned on top of the cylinder head.

Oxides of nitrogen Pollutants formed when nitrogen and oxygen come in contact with combustion chamber temperatures in excess of 2,500°F (1,371°C).

Oxygen A gas making up approximately 18 percent of the air in the atmosphere and required for the combustion process to take place in an engine.

Oxygen sensor A computer system sensor which monitors the oxygen content in the exhaust gas. The signal from the oxygen sensor tells the computer whether a rich or lean condition exists in the air/fuel ratio entering the engine.

Packing slip A list of items and quantities accompanying a shipment.

Paid out A form or slip used to record such cash register transactions as returns or refunds.

Parking brake A mechanical brake on the vehicle used for parking or emergency stopping situations.

Parts specialist A person who is trained and skilled in selling parts at a parts counter.

Percentage of profit Profit accruing on the basis of the selling price of the product.

Perpetual inventory A method of keeping a continuous record of stock on hand through sales receipts or invoices.

Physical inventory Determining stock on hand through an actual count of items.

Pilferage Theft; shoplifting.

Pilot bearing A bearing mounted in the center of the crankshaft which supports the end of the transmission input shaft.

Pinion yoke The yoke mounted on the end of the pinion gear of the differential. The pinion yoke transfers torque from the driveshaft to the pinion gear.

Point-of-purchase (POP) materials Advertising materials used to promote sales.

Point-of-sale (POS) materials Advertising materials such as window banners, placards, and counter cards used at a place of business.

Policy A guiding rule for the conduct of a business.

Port fuel injection An automotive fuel delivery system which has a fuel injector for each cylinder positioned in the intake manifold at the base of the intake valve.

Positive crankcase ventilation (PCV) An emissions system which draws hydrocarbons from the engine crankcase and routes them through the intake manifold to be burned in the engine.

Price leader An item advertised at a very low price to attract customers into a store.

Price sheets Price lists usually accompanying catalogs from manufacturers or warehouse distributors; often in different colors for different types of customers.

Professional One who performs a specialized service as a means of employment, such as an ASE-certified automotive technician.

Profit The amount received for goods or services above the amount of expenses.

Public relations Providing the media with stories of interest worthy of being reported in the news.

Purchase order A form from a buyer that specifies items desired, conditions of sale, and terms of delivery.

Pushrod A valve train component used in engines to connect the lifter to the rocker arm.

R-12 The trade name for a refrigerant commonly used in automotive air conditioning systems. It has been phased out because of its hazard to the earth's ozone layer. R-12 is being replaced by R-134a as an automotive refrigerant.

R-134a The trade name for a refrigerant currently used in automotive air conditioning systems. It is the replacement of choice for R-12.

Rack-and-pinion A steering system which uses a horizontal rack with gear teeth and a pinion gear attached to the end of a steering shaft.

Radiator The device mounted at the front of the vehicle which is used to cool the engine. The hot coolant flows through the radiator. The radiator fins contain the coolant, and air flowing past the fins will remove heat.

Rain check A coupon that guarantees a customer the advertised price of a sold out item when it is again available.

Rebuilt A remanufactured part or assembly.

Reciprocating ball A steering system which uses a worm gear filled with ball bearings to reduce steering effort.

Rectifier An electrical device which converts alternating current (AC) into direct current (DC). The AC current produced by the alternator must be converted to DC in order to charge the battery.

Refrigerant A substance used to carry heat away from an object. The term usually refers to the chemical used in air conditioning and refrigeration systems.

Related items The group of items and tools—such as oil, filter, drain pan, filter wrench, and pour spout—needed to perform a particular job on a car.

Relay An electrical device which allows the remote control of a switch. It normally allows the control of a large current with a much smaller current.

Remanufactured part A part which has been reconditioned to original standards.

Reserve stock Merchandise usually kept in the storeroom to restock shelves and displays in the display area.

Resonator A type of secondary muffler.

Restocking fee The fee charged by the store or supplier for having to handle a returned part.

Retail Selling merchandise to walk-in trade, the do-it-yourselfers.

Return policy The policy established by the store regarding the return of unwanted or unneeded parts. Return policies may include restocking fees or prohibit the return of certain types of parts, such as electrical or electronic components.

Returns Merchandise returned by customers, usually for a refund or exchange.

Right-hand rule Store layout designed to permit customers to move naturally to the right on entering.

Rod bearing caps Rod bearing caps hold the rod to the crankshaft.

Rotor A part of the brake system which turns with the wheel spindle and is clamped by the brake pads for braking action. It is a term that also refers to the rotating part of an alternator that contains the field windings.

SAE An abbreviation for Society of Automotive Engineers, which sets the standards for many products.

Salvage That part or parts of an item retrieved from total loss or scrap which is still suitable for restoration, use, or resale.

Seasonal item An item which appeals to the customer during only a certain part of the year. Ice scrapers and lawn equipment are examples of seasonal items.

Selling price The price at which merchandise is sold, but might differ according to the type of customer (wholesale or retail).

Selling up Selling a customer a better quality item when a lower priced item obviously will not do the job he or she expects of it.

Service bulletin A bulletin which provides supplemental and update information for service manuals.

Service manual A manual which contains the diagnosis and repair procedures for vehicle systems.

Servo The hydraulic piston which applies the bands found in an automatic transmission.

Shelf talkers Small placards or flags placed with the promoted items on display shelves to call attention to sale prices, seasonal items, or new products.

Shoplift The act of stealing goods or display items from a store.

Solvent A chemical which is able to dissolve another substance. Solvents are normally used in the automotive industry as cleaning agents.

Spark plug The device which ignites the air/fuel mixture in the combustion chamber.

Special order An order placed whenever a customer purchases an item not normally kept in stock.

Spot A radio or television commercial.

Stand-up A tall, self-supporting point-of-sale piece, usually flat and printed.

Staple item Items such as oil, coolant, additives, and car care products stocked in the display area that customers regularly purchase.

Statement The bill, usually monthly, that a merchant receives for purchases, and also the bill a merchant sends to credit customers.

Stock order A process by which the store orders more stock from the suppliers.

Stock rotation Selling the older stock on hand before selling the newer stock.

Stud A fastening device which resembles a bolt. It has threads at each end and no head.

Sulfuric acid The acid used as an electrolyte in automotive batteries. It is corrosive and produces explosive hydrogen gas while being charged.

Supersession bulletin A bulletin sent by the parts supplier which lists part numbers that now supersede (replace) previous part numbers.

Supplements Catalog and price sheet changes used to keep items and prices current until new catalogs are issued.

Tabloid brochure A colorful, printed advertising piece used in retail promotions as a newspaper insert, direct mail piece, and for door-to-door delivery.

Technical bulletin A bulletin which explains unusual installation, fit, or maintenance problems associated with a part.

Thermostat (1) The device which controls the water flow through the radiator. It prevents water from flowing through the radiator until the engine is at operating temperature. (2) The device that controls the temperature of an air conditioning system.

Throttle body A fuel injection system which places the fuel injectors in a housing at the location previously used by the carburetor.

Throwout bearing A part of the drive train used with manual transmissions. It presses in the pressure plate fingers to release the clutch. It is also known as the clutch-release bearing and is mounted on the clutch release lever.

Thrust angle The alignment of the rear axle to the vehicle.

Tie rod end The movable joint which connects the tie rods to the steering knuckle. Wear in the tie-rod ends will cause excessive tire wear.

Timing belt A belt which connects the camshaft and crankshaft, synchronizing their rotation. Timing belts are normally used with overhead cam engines.

Timing chain A chain which connects the camshaft and the crankshaft, synchronizing their rotation.

Timing gears The gears mounted on the end of the camshaft and crankshaft. They mesh together and synchronize the rotation of the camshaft and crankshaft.

Torque converter The device located between the engine and the automatic transmission. It will slip and allow the engine crankshaft and automatic transmission input shaft to rotate at different speeds.

Torque multiplication The process of increasing the torque output of a torque converter. Torque converters usually provide some amount of torque multiplication for vehicle acceleration.

Trade discount A discount from the regular selling price offered to high-volume purchasers such as dealers, fleet owners, and industrial companies.

Traffic Customers.

Traffic builders All types of advertising, promotions, and merchandising materials designed to bring customers into a store.

Transaxle A combination transmission and differential.

Transfer case A gearbox which connects the front and rear drivelines of a four-wheel drive vehicle.

Transmission A set of gears which can change ratios to meet the various driving needs of the vehicle.

Transverse An orientation which refers to the engine and transaxle being mounted crosswise in the vehicle.

Trim tag A tag located on the vehicle body which outlines the color, body style, trim, and body accessories found on the vehicle.

Turbocharger A device mounted on the engine which forces air into the intake manifold. Turbochargers are driven by exhaust gas from the engine and serve to boost power and performance.

Turnover The number of times each year that a merchant buys, sells, and replaces a part number.

Universal joint (U-joint) A flexible coupling which connects the driveshaft to the pinion yoke of the differential.

Vehicle identification number (VIN) A unique number assigned to a vehicle for identification purposes. The VIN number can be decoded for information regarding year of manufacture, manufacturer, body style, engine size, carrying capacity, and other information.

Vendor The supplier.

Voltage regulator The charging system device which sets the maximum charging voltage produced by the alternator.

Walk-in Retail and do-it-yourself customers.

Warehouse distributor (WD) The jobber's supplier. The link between manufacturer and jobber.

Warranty A printed document stating that a product will provide satisfactory service for a given period of time or the buyer will be entitled to a settlement according to the terms of the warranty.

Warranty return A defective part returned to the supplier due to failure during its warranty period.

Water pump The pump located at the front of the engine which circulates engine coolant throughout the cooling system.

WATTS Long-distance telephone service used by companies making a large number of long-distance calls daily.

Wheel bearings The bearings located between the wheel hubs and spindle or axle housing and the axle.

Wholesale The merchant's price to large volume customers such as dealers, fleet owners, and industrial companies. Unlike a trade discount, it is usually a fixed price rather than a percentage off the retail price.

Will-call Merchandise held for a customer to pick up.

Window banners Large posters displayed in windows to attract passersby to items on sale or promotions in progress.

Wire hangers Point-of-sale advertising materials attached to or draped over a wire inside the store.

Worm gear A gear cut in such a way that it spirals around a shaft.

Notes

Notes

Notes